FORECASTING

DEMAND AND SUPPLY OF

DOCTORAL SCIENTISTS

AND ENGINEERS

Report of a Workshop on Methodology

Office of Scientific and Engineering Personnel
National Research Council

NATIONAL ACADEMY PRESS
Washington, D.C.

NATIONAL ACADEMY PRESS • 2101 Constitution Ave., NW • Washington, DC 20418

NOTICE: The project that is the subject of this report was approved by the Governing Board of the National Research Council, whose members are drawn from the councils of the National Academy of Sciences, the National Academy of Engineering, and the Institute of Medicine. The members of the committee responsible for the report were chosen for their special competencies and with regard for appropriate balance.

Support for this project was provided by the National Science Foundation and the Sloan Foundation. Opinions, findings, conclusions, or recommendations expressed in this publication are those of the authors and do not necessarily reflect the views of the sponsors.

International Standard Book Number 0-309-07089-9

Additional copies of this report are available from:
National Academy Press (http://www.nap.edu)
2101 Constitution Avenue, N.W., Box 285
Washington, D.C. 20055
800-624-6242
202-334-3313 (in the Washington metropolitan area)

THE NATIONAL ACADEMIES

National Academy of Sciences
National Academy of Engineering
Institute of Medicine
National Research Council

The **National Academy of Sciences** is a private, nonprofit, self-perpetuating society of distinguished scholars engaged in scientific and engineering research, dedicated to the furtherance of science and technology and to their use for the general welfare. Upon the authority of the charter granted to it by the Congress in 1863, the Academy has a mandate that requires it to advise the federal government on scientific and technical matters. Dr. Bruce M. Alberts is president of the National Academy of Sciences.

The **National Academy of Engineering** was established in 1964, under the charter of the National Academy of Sciences, as a parallel organization of outstanding engineers. It is autonomous in its administration and in the selection of its members, sharing with the National Academy of Sciences the responsibility for advising the federal government. The National Academy of Engineering also sponsors engineering programs aimed at meeting national needs, encourages education and research, and recognizes the superior achievements of engineers. Dr. William A. Wulf is president of the National Academy of Engineering.

The **Institute of Medicine** was established in 1970 by the National Academy of Sciences to secure the services of eminent members of appropriate professions in the examination of policy matters pertaining to the health of the public. The Institute acts under the responsibility given to the National Academy of Sciences by its congressional charter to be an adviser to the federal government and, upon its own initiative, to identify issues of medical care, research, and education. Dr. Kenneth I. Shine is president of the Institute of Medicine.

The **National Research Council** was organized by the National Academy of Sciences in 1916 to associate the broad community of science and technology with the Academy's purposes of furthering knowledge and advising the federal government. Functioning in accordance with general policies determined by the Academy, the Council has become the principal operating agency of both the National Academy of Sciences and the National Academy of Engineering in providing services to the government, the public, and the scientific and engineering communities. The Council is administered jointly by both Academies and the Institute of Medicine. Dr. Bruce M. Alberts and Dr. William A. Wulf are chairman and vice chairman, respectively, of the National Research Council.

OFFICE OF SCIENTIFIC AND ENGINEERING PERSONNEL
ADVISORY COMMITTEE

1997 – 98 Membership

M. R. C. Greenwood, University of California, Santa Cruz, *Chair*
David Breneman, University of Virginia
Nancy Cantor, University of Michigan
Carlos Gutierrez, California State University, Los Angeles
Stephen J. Lukasik, Independent Consultant
Barry Munitz, J. Paul Getty Trust
Janet Norwood, The Urban Institute
John D. Wiley, University of Wisconsin, Madison
Tadataka Yamada, SmithKline Beecham Corporation
Thomas Young, Lockeed Martin Corporation (Retired)

Ex-officio Member

William H. Miller, University of California

Staff

Charlotte Kuh, *Executive Director*
Marilyn Baker, *Deputy Executive Director*

PREFACE

his report presents the findings of a workshop organized by the
Committee on Methods of Forecasting Demand and Supply of
Doctoral Scientists and Engineers. The committee examined the
methodologies that have been used to forecast labor market
conditions for scientists and engineers, identified and analyzed
additional methodologies, and recommended improvements in the
way that forecasts are presented to users. The committee was
charged with three tasks. These were:

1. Identify how estimates of supply and demand for
scientists and engineers are currently used by policymakers with
particular emphasis on how these estimates relate to the level of
federal funding for research and development and the financial
health of the academic sector.

2. Identify the sources of uncertainty, outline the importance of definitions and underlying assumptions to state-of-the-art
projection methodology, and discuss how these uncertainties and
assumptions can be fairly represented so that policymakers can
understand both the strengths and limitations of the estimates that
they see.

3. Recommend ways of presenting projections so that the
sources of uncertainty are explicitly taken into account.

The *Workshop on Improving Models of Forecasting Demand and Supply
for Doctoral Scientists and Engineers* met on March 19–20, 1998. Four

panels of experts were invited to participate in presentations and discussions. Biographical descriptions of these participants are presented in Appendix B. The methodological issues focused on four areas:

1. **Forecasting Models: Objectives and Approaches** identified the characteristics of good models of supply and demand for scientists and engineers. The panel discussed how adjustment can occur in three dimensions: quality, price, and quantity. The session focused on alternative approaches to these models and described what models could be estimated given available data.

2. **Neglected Margins: Substitution and Quality** examined the effects of price, substitution, and immigration on the modeling of labor markets for scientists and engineers.

3. **Models of Scientific and Engineering Supply and Demand: History and Problems** focused on shortage/surplus or "gap" models used for estimation in the past. The panel examined their usefulness, purposes, and their potential for modification of gap models in order to take into account the simultaneous adjustment of quality and quantity.

4. **Presentation of Uncertainty and Use of Forecasts with Explicit Uncertainty** investigated the best ways to communicate the sensitivity of model outputs to assumptions and uncertainty. The panel focused on how uncertainty should be presented to policymakers and others who are educated but not expert users.

Forecast users and others with an interest in science personnel policy were invited to discuss their concerns about forecasts and their use. Appendix C lists persons who attended the workshop. Following the workshop, the committee met to formulate recommendations about productive avenues for research, data, and/or dissemination of the results of models for forecasting the demand and supply for doctoral scientists and engineers. The committee reached consensus on five recommendations, which appear in this report.

M. R. C. Greenwood
Chair
Office of Scientific and Engineering
Personnel Advisory Committee

ACKNOWLEDGMENTS

The *Workshop on Improving Models of Forecasting Demand and Supply for Doctoral Scientists and Engineers* benefited from the contributions of many people and was funded by the National Science Foundation (NSF) and the Sloan Foundation. The Committee on Methods of Forecasting Demand and Supply of Doctoral Scientists and Engineers acknowledges those who made the workshop successful. First and foremost are those workshop participants who prepared manuscripts that framed the issues for each panel. They include:

- Bert S. Barnow, *Objectives and Approaches of Forecasting Models for Scientists and Engineers*
- George Johnson, *How Useful Are Shortage/Surplus Models of the Labor Market for Scientists and Engineers?*
- Sherwin Rosen (and Jaewoo Ryoo), *The Engineering Labor Market*
- Nancy Kirkendall, *All Models Are Wrong; Some Models are Useful.*

Special appreciation is expressed to Michael McGeary, who prepared Chapter 1 of this report. I wish to thank the members of the committee for their contributions to the workshop. In addition, important contributions to the workshop's success were made by the workshop participants: Michael Teitelbaum, Alfred P. Sloan Foundation; Jeanne Griffith, National Science Foundation; John A. Armstrong, IBM; Ronald Ehrenberg, Cornell University; Michael Finn, Oak Ridge Institute for Science and Education; Paula

Stephan, Georgia State University; Eric Weinstein, MIT; Geoff Davis, Dartmouth College; Charles A. Goldman, RAND; Sarah E. Turner, University of Virginia; Robert Dauffenbach, University of Oklahoma; Daniel Greenberg, Johns Hopkins University; Neil Rosenthal, Bureau of Labor Statistics; Alexander H. Flax, Institute for Defense Analyses; and Skip Stiles, Committee on Science, U.S. House of Representatives.

This report has been reviewed by persons chosen for their diverse perspectives and expertise in accordance with procedures approved by the National Research Council's Report Review Committee. This independent review seeks to provide candid and critical comments that will assist the Office of Scientific and Engineering Personnel (OSEP) in making its report as sound as possible and will ensure that the report meets institutional standards of objectivity, evidence, and responsiveness to the study charge. The review comments and draft manuscript remain confidential to protect the integrity of the deliberative process. We wish to thank the following for their participation in this review of the report: John Armstrong, Erich Bloch, and Robert Lerman.

The project was aided by the invaluable help of the OSEP professional staff—Charlotte Kuh, executive director; George R. Reinhart, project officer; Cathy Jackson, administrative associate; and Margaret Petrochenkov, who provided editorial input.

Finally, we wish to express our gratitude *in memoriam* to Alan Fechter, former executive director of the Office of Scientific and Engineering Personnel. Forecasting demand and supply for scientists and engineers was one of Alan's ongoing concerns, and this report is dedicated to his memory.

Daniel McFadden
Chair
Committee on Methods of Forecasting
Demand and Supply of Scientists and
Engineers

CONTENTS

EXECUTIVE SUMMARY

Background

Interest in predicting demand and supply for doctoral scientists and engineers began in the 1950s, and since that time there have been repeated efforts to forecast impending shortages or surpluses. As the importance of science and engineering has increased in relation to the American economy, so has the need for indicators of the adequacy of future demand and supply for scientific and engineering personnel. This need, however, has not been met by data-based forecasting models, and accurate forecasts have not been produced.

Forecast error may proceed from many sources. Models may be based on incorrect assumptions about overall structure, included variables, lag structure, and error structure. Data used for estimation may be flawed or aggregated at an inappropriate level. Further, unanticipated events beyond those considered in the model may occur that could ruin the accuracy of even the best forecasts. As Leslie and Oaxaca (1993) have described in their thorough review article, virtually all models of demand and supply have been flawed by at least one (and, in many cases, all) of these problems.

In order to assess the methodology of forecasting the demand and supply of doctoral scientists and engineers, the National Science Foundation (NSF) and Sloan Foundation funded the National Research Council to assemble a committee of experts for a workshop on the topic. The task of the committee was not to find fault with past efforts, but to provide guidance to the NSF and to scholars in this area about how models (and the forecasts derived from them) might be improved and what role NSF should play in their improvement.

Another issue for the committee was the responsible reporting of forecasts to policymakers. Virtually no forecast is error-free, and some uncertainty is always associated with them. Policymakers who use the results of forecasting are usually not technically proficient in the arcana of the forecaster's art. If forecasts are to be used responsibly, policymakers need to be informed about the assumptions upon which the forecasts rest, and forecasters need to track the validity of their assumptions and the accuracy of their forecasts over time.

The committee then assessed the information on these issues provided at the workshop as well as in the forecasting literature and arrived at the following recommendations.

Recommendations

The forecasting process is ongoing. Forecasters must learn from their mistakes. The whole forecasting exercise needs to be placed within an administrative framework that facilitates an evaluation process and a process to correct errors. The Science Resources Study Division (SRS) of NSF is the locus of action in the federal government to bring about the improvement of data and forecasts of the market for doctoral scientists and engineers. The committee has therefore directed most of its recommendations to SRS, since this division should be able to encourage improvements in the construction and use of forecasting models for highly trained scientists and engineers, even when SRS should not carry out the work itself.

Recommendation 1. The producers of forecasts should take into account the variety of consumers of forecasts of demand and supply for scientists and engineers.

NSF should recognize that there are five distinct communities of clients for data and projections on the supply and demand of scientists and engineers, each with different needs and interests. NSF's data collection and forecasting activities should keep the needs of these different communities in mind. They are:

1. students making career decisions,
2. federal, state, and private funding agencies,
3. industrial and academic employers of scientists,
4. Congress in its role as funder and policymaker, and
5. the scientific community that conducts research and produces studies of the market and its participants.

Representatives of all these user communities spoke at the workshop. All expressed dissatisfaction with the current state of data and forecasting.

Students need qualitative projections of likely career outcomes and probabilities of success, with particular attention to the state of the job market in a few years when they will seek employment. These qualitative projections require timely data, but they need not be based on broad surveys or censuses.

Research funding agencies and Congress need relatively long-term projections of supply and demand factors by specific discipline that can be used to guide policy on training support and institutional development. These projections should take careful account of the ease with which one kind of labor can be substituted for another and the incentives and behavioral responses that operate. The importance of contingencies and of forecast uncertainty should be emphasized.

The needs of employers are varied, but they also benefit from an early warning about shortages. As producers of Ph.D.s, academic institutions can take steps to expand or contract Ph.D. enrollment given convincing evidence of emerging labor market trends.

Finally, the forecasting community should be more forthcoming about the appropriate use of forecasts and the nature of their underlying assumptions.

Recommendation 2. The NSF should not produce or sponsor "official" forecasts of supply and demand of scientists and engineers, but should support scholarship to improve the quality of underlying data and methodology.

NSF should not produce or sponsor "official" forecasts of supply and demand in the markets for scientists and engineers, but should continue to take the lead in collecting and making available data on these markets. A clear organizational separation should be made between data collection and modeling/forecasting activities undertaken for NSF's own policy use or for use by federal agencies. For example, convert the SRS into a National Center for Science Statistics on the model of the National Center for Health Statistics (NCHS), the National Center for Educational Statistics (NCES), or the Energy Information Agency (EIA) and remove modeling and forecasting activities to a separate policy unit. Or, for example, forecasts could be produced by an outside agency with statistical expertise. Agencies such as the Bureau of the Census or Bureau of Economic Analysis may be well suited to undertake such forecasts.

If asked to produce forecasts of scientific and engineering personnel for its own use or the use of other agencies, the NSF policy unit should avoid endorsing or emphasizing "gap" models that do not incorporate behavioral adjustment to demand and supply and consequently may give unwary users a misleading impression of likely market outcomes. NSF should avoid suggesting that there is a single best level of detail and model complexity for the forecasts needed by various users and should instead maintain that model structure will depend on user needs and objectives.

The committee reviewed the history of NSF projections, most notably those in *The State of Academic Science and Engineering*.

In response to the perception that there could be a connection between NSF funding and its projections, especially regarding projections of shortages, the committee believed that NSF should limit itself to data collection and dissemination, and use external "arms-length" forecasts for its policy needs to avoid the possible conflict of interest that might occur if it produced its own forecasts.

Recommendation 3: Undertake a comprehensive review of data collection in the light of forecasting needs.

NSF's SRS should undertake a comprehensive review of its data management program, preferably in coordination with Bureau of Labor Statistics (BLS), NCHS, and NCES. It should seek the production of more timely and useful data on the market for scientists and engineers. In addition, it should coordinate definitions and categories across agencies to facilitate a consistent picture of the different stages in the market, from student training and degree choice to mid-career transitions across and out of science and engineering fields. Moreover, sample sizes have been reduced since the late 1970s, which makes modeling difficult for small fields, specific employment sectors, and for rare events (such as mid-career changes).

Recommendation 4. Data that enhance forecasts should be widely available and be disseminated on a timely basis.

NSF's SRS should establish three high-priority data objectives. These are: (1) production of timely, descriptive statistics on employment and salaries by field of training, occupation, and sector, in a consistent time-series format that permits tracking and projections of trends; (2) production of an individual-level Public Use Sample, containing a consistent time series of cross-sections of doctoral recipients, and when available, nondoctoral recipients; and (3) production of a Public Use dataset panel of scientists and engineers to analyze transitions from the educational system to employment and transitions across fields, occupations, and activities. NSF should process these data, establish rules for access so that they are widely available, and institute less burdensome mechanisms to protect confidentiality. For example, in producing a Public Use panel, NSF might recruit respondents who are willing to provide *vitae* in a standard format without confidentiality restrictions. This format could be made available on the web; to make the process even easier for respondents, NSF could code respondent *vitae*. This would enhance modeling of individual and institutional behavior in response to changes in funding, demographically driven demand, compensation, etc. Standard coding of *vitae* would permit collection of more detailed data than the *Survey of Doctorate Recipients* form.

Both modelers and policy advisors at the workshop complained that data were not timely. At present, there is no way of knowing in a timely way whether market mechanisms are working to alleviate shortages or gluts. Policies calibrated to outdated evidence may provide too little too late, or too much too late. Furthermore, if indicators show that market-clearing mechanisms are working rapidly, the need for policy change might be obviated altogether.

NFS's SRS should collect and disseminate data that enable a variety of forecasting exercises that differ along some or all of the following dimensions:

    a. Unconditional vs. conditional ("What if") forecasts.

    b. Multiple levels of disaggregation by field, e.g., physical sciences, physics, solid state physics, materials science.

    c. Optional variables to forecast, e.g., jobs, salaries, quality, productivity, occupation.

    d. Various sectors to forecast, e.g., academic tenure track/ other, postdoctoral positions, or industrial sectors.

    e. What to forecast, e.g., stocks, flows, transitions, or careers.

    f. Permit forecasts to go beyond means and standard errors

to more complete descriptions of the distribution of possible outcomes.

g. Various time horizons for forecasts.

Finally, the NSF's SRS website currently focuses almost exclusively on providing relatively simple tabulations that are useful for casual policy analysis but not very useful for either career planning by students or for research on the science and engineering market. The data management program and website should be redesigned to service these neglected user communities (or in the case of students, the public and private organizations and associations that provide career guidance), and to provide links to other data sources from BLS and NCES that are important for analysis of the markets for scientists and engineers.

Recommendation 5. NSF should develop a research program to improve forecasting.

The Directorate of Social, Behavioral and Economic Sciences of the NSF should commission behavioral studies of scientists and engineers early in their careers, as well as studies of forecasting methods and evaluations of past forecasting exercises. Particular emphasis should be placed on critical parameters of market response, such as wage-sensitivity of field and occupation switching, and the determinants of the ability of employers to restructure research jobs in response to supply conditions in various fields. Emphasis should be placed on the difficult issues of measuring the quality and productivity of scientists and engineers, the quality of worker-to-job matches, and quality of life for scientists and engineers. This work should be conducted through the ordinary peer-reviewed research support process already in place at NSF. SRS should facilitate the dissemination of results of these studies but should stop short of sponsoring or endorsing specific forecasts or methods.

CHAPTER 1

BACKGROUND

Issues in the Prediction of Demand and Supply for Doctoral Scientists and Engineers

Interest in predicting demand and supply for doctoral scientists and engineers began in the 1950s. Arrow and Capron (1959) found the academic labor market interesting, in part because it was characterized by lengthy training periods for Ph.D.s. This meant that the market students saw when they began their graduate training was often very different from the market they encountered when they finished. These sorts of lags led to alternating spells of oversupply and undersupply (often called cobweb adjustment) in some fields. Freeman (1971) and Cain et al. (1973) carried out empirical explorations of these models in the 1970s. At the same time, Cartter (1970, 1976) made demographic models of the demand for faculty and forecast an end to the booming market for Ph.D.s in the late 1960s and early 1970s, which, in part, had been driven by the growing demand for faculty to teach the baby boom generation of students.

In addition to the surging demographic demand for doctoral scientists and engineers, the late 1950s through the late 1960s saw a tremendous increase in federal funding for research in the academic, industrial, and government sectors, which added additional fuel to demand and a growing recognition that this sector of the work force is critical in meeting national needs. The same period saw steep increase of graduate fellowships and traineeships supported by NSF and federal agencies. In particular, the National Aeronautics and Space Administration (NASA), the Department of Defense (DOD), and the National Institutes of Health (NIH) funded predoctoral and postdoctoral fellowships and training grants in the biomedical and behavioral sciences. As a result, the number of science and engineering Ph.D.s awarded annually tripled, from 6,000 in 1960 to more than 19,000 in 1971. Federal programs for doctoral training were closed down or sharply reduced in the early 1970s, when demographic trends, reductions in federal R&D funding, and unemployment among new Ph.D.s indicated an oversupply condition. However, Congress kept the supply of biomedical researchers up by passing a special law, the National Research Scientist Act (NRSA), and overseeing its implementation. Congress believed that NIH research training should be related to market demand and mandated the National Research Council to advise NIH on the nation's need for biomedical and behavioral research personnel for the next 10 years. The National Research Council (NRC) has done so in biennial and quadrennial reports since 1976 (NRC, 1994).

Consequently, increases in the annual number of science and engineering Ph.D. degrees ended, and the number declined slightly after 1971 before starting to grow again in 1979. The 1972 total was not surpassed until 1986. Also during the 1980s and into the 1990s, foreign-born graduates with temporary visas accounted for almost all the growth in the annual number of Ph.D.s. Although some investigators expressed skepticism (Ehrenberg, 1991), in the late 1980s the NSF and others estimated that there might not be enough new science and engineering faculty to replace those hired to teach the children of the baby boom (NSF, 1989; Bowen and Sosa, 1989; Atkinson, 1990). To date, quite the opposite has occurred. In response to a sharp buildup in military research and development during the first half of the 1980s and widely publi-cized projections of a shortage of doctoral scientists and engineers, the number of Ph.D.s granted increased steadily until a year or two ago, due in part to an increase in noncitizen doctorates, as shown in Figure 1-1. By the early 1990s, a growing share of new Ph.D.s was experiencing difficulty in obtaining permanent employment upon graduation. In the biological sciences, this has led to a growing pool of individuals in postdoctoral positions (NAS, 1998). Nonacademic

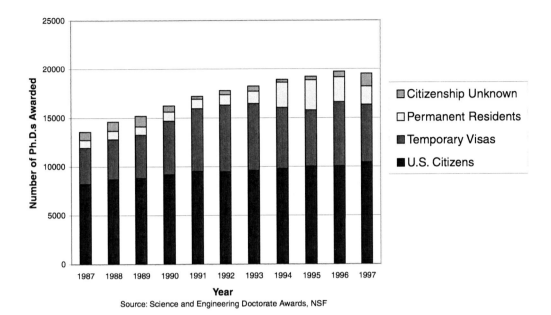

FIGURE 1-1 *Number of natural science and engineering Ph.D. awards by citizenship and year.*

employment now accounts for over half of the employment of Ph.D.s in most scientific and engineering fields (NAS, 1993). Government demand for new Ph.D.s has been declining. Academic hiring has remained flat because state contributions to public colleges and universities have declined, and academia, both public and private, has come under increasing pressure to slow the growth of tuition and costs. Industry demand, which is cyclically sensitive, has been growing slightly, but many of the large industrial laboratories have been drastically downsized in the past decade.

These forecasts of undersupply that did not materialize have led policymakers for graduate training and research support to be highly skeptical of any forecasts and to worry about the self-interest of the forecasters. Models that have predicted an extreme oversupply have received favorable media attention (Massy and Goldman, 1995), and such models have sharpened concern in policy circles that the federal funding of graduate education and research may simply be aggravating the oversupply. Conflicting forecasts have cast further doubt in the public mind on the usefulness of any forecasting at all. Still, the role of the nation's human resources in science and engineering remains critical. Against this background, the NSF and the Sloan Foundation asked the NRC to form a committee of model makers and users to assess the current state of the art of models that are used for forecasts, to suggest directions for improvement, and to assist policymakers in the informed use of forecasting models.

Issues for the Committee

Forecast error may proceed from many sources. Models may be misspecified with respect to overall structure, included variables, lag structure, and error structure. Data used for estimation may be flawed and aggregated at an inappropriate level. Further, unanticipated outside events may occur that can ruin the accuracy of even the best of forecasts. As Leslie and Oaxaca (1993) have described in their thorough review article, virtually all models of demand and supply have been flawed by at least one (and, in many cases, all) of these problems. The task of the committee was not to find fault with past efforts, but to provide guidance to NSF and to scholars in this area about to how models (and the forecasts derived from them) might be improved and what role NSF should play in their improvement.

Another issue for the committee was the responsible reporting of forecasts to policymakers. Virtually no forecast is error-free, and some uncertainty is always associated with it. Policymakers who use the results of forecasting are not usually technically proficient in the arcana of the forecaster's art. If forecasts are to be used responsibly, policymakers need to be informed

about the assumptions upon which the forecasts rest, and forecasters need to track the validity of their assumptions and the accuracy
of their forecasts over time. For example, if there appears to be a
permanent decline in state funding of higher education, forecasters
need to recast their models to reflect these changed conditions and
must inform those who use their models about the implications of
the change. Other audiences to whom this report is directed
include graduate students in science and engineering and employers of these students—including industrial and academic employers, both federal and private funding agencies, and the scientific
community itself. How best to inform policymakers who would be
interested in using forecasts of demand and supply of scientists and
engineers is also an issue that needs to be addressed.

In order to learn from both forecast makers and forecast
users about improvements that can be made in understanding the
markets for doctoral scientists and engineers, the committee
sponsored a workshop at NAS on March 20 and 21, 1998. Papers
were commissioned on: (1) the history and problems with models of
demand and supply for scientists and engineers, (2) objectives and
approaches to forecasting models, (3) margins of adjustment that
have been neglected in models, especially substitution and quality,
(4) the presentation of uncertainty, and (5) whether forecasts of
supply and demand for scientists and engineers are worthwhile,
given all their shortcomings. The workshop discussion is summarized in Chapters 2-5. (The agenda and list of participants for the
workshop appears in Appendixes A and B.) Following the workshop, the committee met and agreed to a set of recommendations
that are presented in Chapter 6.

CHAPTER 2

MODELS OF THE DEMAND FOR DOCTORAL SCIENTISTS AND ENGINEERS: HISTORY AND PROBLEMS

It is important to the nation that there be an adequate number of scientists and engineers. Industries that rely on scientific and technological research and development are increasingly important in both the global and the American economies. If there are too few scientists and engineers, the economy and its competitive position, both now and in the future, are put at risk. An adequate supply of doctoral scientists and engineers is also important to the nation's colleges and universities, since these institutions train both graduate students and undergraduates and carry out university research.

Conversely, if too many people trained as scientists and engineers cannot find work related to their training, costs also result to both the scientists and engineers, and to the federal government, state governments, and universities that have subsidized the many years of education that cannot be used appropriately. In the late 1980s, predictions that shortages of doctorates would emerge in the 1990s appeared not only in the technical literature (Bowen and Sosa, 1989; NSF, 1989), but also in the media (Sovern, 1989). Young people who went to graduate school at that time and expected a welcoming job market when they received their degrees were, in many cases, sorely disappointed.

A simple illustration of the way that economists analyze the costs of shortages or surpluses is shown in the four boxes in Figure 2-1.

FIGURE 2-1   *The social and individual costs of shortage and surplus.*

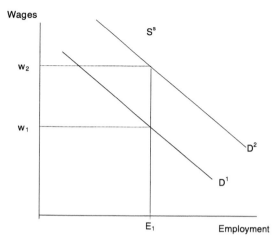

Box 1: In the very short run

When labor demand increases from D^1 to D^2, perhaps because of the implementation of a new technology, the supply of scientists is fixed at E_1 and the increase in demand only increases the wages of scientists currently in the labor force from w_1 to w_2. The wage bill increases from $w_1 E_1$ to $w_2 E_1$. If wage adjustment is not immediate, there will be a short-run gap, met by rationing workers and postponing tasks.

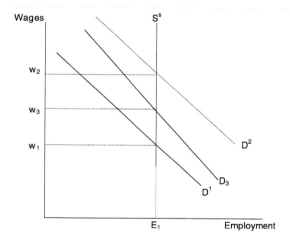

Box 2: In the short run

Employers, seeing the jump in the wage needed to attract skilled workers, try to find ways to use these workers more efficiently. They might, for example, hire more technicians or programmers so that Ph.D. scientists can focus on research, transfer scientists from related fields, or adjust qualifications. Their demand curve shifts from D^2 to D^3, and their wage bill, $w_3 E_1$, is lower than it would have been without this change.

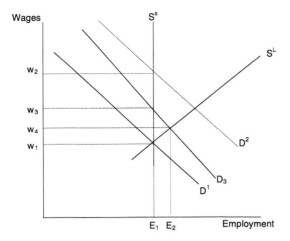

Box 3: In the longer run
Students see the increase in wages and move into the field. At the higher wage, more Ph.D.s are available. The supply curve rotates to S^L. The increased supply means that employers pay lower wages, w_4, and that more workers, E_2, are available than in the absence of the supply adjustment.

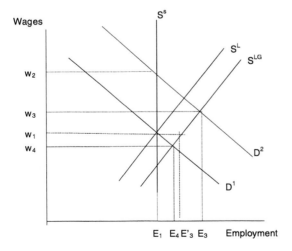

Box 4: Government policy
If the government anticipated the increase in demand, it could try to increase supply in advance to S^{LG} by providing funding for graduate students. This would shift the supply curve, resulting in more skilled workers, E_3 and lower wages, w_3, than if there were no policy. If, however, the increase in demand fails to materialize, the result would be unemployment, which could be as large as $E'_3 - E_1$ if wages fail to adjust, or lower wages, w_4, or some combination of the two.

The need to anticipate future demand for doctoral scientists and engineers, and to adjust factors that affect the supply to meet the demand, has encouraged the construction of forecasting models. Yet, as Johnson (1998) pointed out in his workshop paper, forecasting these markets beyond the very short term is extraordinarily difficult:

> The future paths of levels of employment and wages in professional labor markets depend on the paths of at least ten exogenous variables. . . . Two or three of these exogenous variables, dealing primarily with the age structure of the population, can be predicted quite well for twenty or more years into the future. Other important exogenous variables, like the distribution of preferences of future college students for scientific versus other careers and future technological changes that might affect the demand for scientific personnel are inherently unpredictable. Further, there are important substantive issues involving, among other things, the heterogeneity of labor input within scientific fields and the substitution among groups of inputs with different quality/qualifications. These factors could drastically affect market forecasts and are not satisfactorily understood.

Even apparently predictable demographic change can become unpredictable if immigration increases.

Essentially, good forecasts require a well-specified model that correctly reflects behavior, is based on good data as well as careful forecasts of variables determined outside the model, and is produced by parties with no vested interest in actions taken in response to the forecast. Although past models, which were exhaustively reviewed in Leslie and Oaxaca (1993), often satisfied one of these requirements,[1] few have been formulated to reflect the interaction of wages, quantities, and quality, let alone what the committee called the "neglected margins" of interchangability of workers across fields, by degree level, quality, or nationality.

What happens to forecasts when models leave things out? Put simply, they can mislead. For example, the NSF models of the late 1980s did not anticipate a deep economic recession and its effect on state and federal education and R&D budgets. Moreover, they did not anticipate the end of the Cold War and its effect on defense spending, which translated into fewer jobs for scientists and engineers. They failed to anticipate the impact of legislation that abolished mandatory retirement, thereby postponing (but not eliminating) the expected retirement of doctoral faculty who had been hired in the late 1960s and early 1970s to teach baby boom children. All of these unanticipated exogenous events worked to dampen the demographically based forecasts of an increase in demand for doctoral scientists and engineers. In addition, the models failed to account for the market mechanisms that operate to bring supply and demand into balance: wage adjustments, immigration, transfer of workers across fields, and adjustments in qualifications. The difference between the NSF forecast of the number of jobs that would be available for new Ph.D.s and the actual number of new Ph.D.s placed in science and engineering jobs is shown in Figure 2-2.

This incorrect forecast resulted from unanticipated changes in the economic environment, exacerbated by the model's neglect of market adjustment mechanisms. As Johnson points out, it is beyond the capacity of a long-term market forecasting model to

---

[1] In particular, good demographic data and forecasts exist, as well as long time series on levels, numbers, and fields of degrees. A number of models have forecast supply and demand for scientists and engineers based on demographic variables and industrial demand based on input/output models of the U.S. economy. These models typically do not include wages and are rarely disaggregated by field.

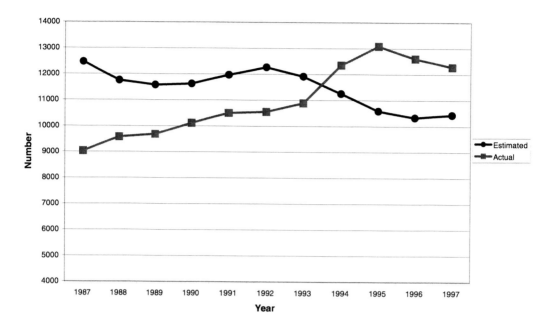

FIGURE 2-2   *The NSF/Atkinson forecast of supply of new Ph.D. scientists and engineers in the early 1990s and what actually happened.*

anticipate all the changes that can affect demand and supply. However, if regularly updated models had been available that took into account recent events and market adjustment mechanisms, and communicated the true limits of forecast accuracy, then forecast errors would have been smaller and less surprising. Such models could have provided an earlier warning that the market outlook would not be as rosy as the analysts of the late 1980s anticipated. Instead, policymakers were surprised by the increase in the numbers of new Ph.D.s who did not have definite job commitments upon graduation and by the rapid growth in the pool of postdoctoral students, especially in the biological sciences. In response, the National Science Board asked the National Research Council to study the question of how to "reshape" graduate education to improve the chances that new Ph.D.s find appropriate employment outside of academia (NAS, 1993). At the same time, Massey and Goldman (1995) constructed a detailed, field-disaggregated simulation model that forecast considerable oversupply in most science and engineering fields. Their report pointed to the demand for graduate students driven by research funding as an important explanation of why production of new Ph.D.s in science and engineering exceeded academic demand. However, wages and other market adjustment mechanisms were also missing from the Massy-Goldman model.

Much of the discussion that followed the Johnson paper focused on how models might be improved. Areas for improvement fell into three broad categories:

- *Data.* Good data exist on Ph.D.s and especially on Ph.D.s in academia. Data on career paths of Ph.D.s in industry (or the classifications that would permit characterization of career paths) are lacking, as are sample sizes large enough to permit study of subpopulations in fine fields (e.g., bioinformatics). Since Ph.D. data are collected for individuals, consistency with data collected by occupations is difficult to achieve. Wage data (but not data on total compensation) by occupation are available, but not by different degree levels. Data are collected for international students who receive degrees from U.S. universities, but retention rates are difficult to determine. Data on immigrants who enter the United States after earning their Ph.D. have been collected only recently. The unavailability of data makes it difficult to model the mobility of scientists and engineers in response to changing market conditions. Finally, very little data allow for the measurement of changes of quality.
- *Model specification.* Most models to date have been "gap" models that project demand and supply separately. These models neglect wage adjustments and other market mechanisms that tend to modify demand and supply and bring them into

balance. Gap models thus describe structural conditions prior to the operation of these mechanisms or describe a world in which these mechanisms are believed to be ineffective. Better models would estimate demand and supply simultaneously and incorporate behavioral parameters so that changes in demand could, over time, feed back to supply and vice versa.

- *Omitted variables.* Most models leave out immigration (in part because good data are lacking) and workers in science with nonscience degrees. Barnow (1998) noted in his paper that for bachelor's degree holders, at least, these flows can be substantial. For those with a bachelor's degree in computer science, 53.5 percent take a job outside computer science, while only one-third of those with computer science jobs received their degree in that field. These flows may be less prevalent in markets for Ph.D.s, but are likely to become more important during periods of increased demand for particular types of workers. A current example is the field of bioinformatics.

OBJECTIVES AND APPROACHES OF FORECASTING MODELS

There are many potential users of models that forecast demand and supply of workers for particular occupations. One group of users includes students, their parents, and advisors. Others include universities that need to shape the size and scope of their graduate programs, industry planners who need to anticipate worker shortages in key areas of emerging technology, and government funders of graduate education and research, at both state and federal levels, who need to allocate public funds wisely. What is desired in a forecasting model may vary for each of these groups, in terms of horizon, level of detail, and focus.

Burt Barnow, in his workshop paper, described these different objectives:

> What we want from the occupational forecasts depends on their use. For career decisions, either by individuals or by training programs on behalf of the participants, we are primarily interested in the employment and pay situation for specific occupations over a fairly long period. For meeting national priorities, however, the number of individuals in a particular occupation is a means to the ends—it may be possible to substitute workers with the right skills but in other occupations to achieve the same ends.

He pointed out that quality, as measured by ability, skill, or credentials, is an important dimension of market adjustment that is omitted from most forecasting models. When faced with shortages, employers can hire less well trained workers, but they can also redefine jobs. It is important to be able to predict whether shortages will persist. Is there a shortage because employers do not wish to pay the wages needed to attract workers of a desired level of quality? Have employers explored substitution possibilities fully? In the long run, would the shortage remain? This might be the case for scientists and engineers if, for example, students lacked the background in mathematics needed to undertake a science or engineering career. In this case, the lag until supply adjusted to meet demand might be quite long, and changes in immigration policy might be worth considering.

These questions, however, simply underscore the need for better data and more careful modeling of market adjustment. These models could be used to address several questions. How long are adjustment lags? By how much might wages rise in the absence of supply change? How rapidly can students and experienced workers change fields, thereby increasing supply?

One encouraging development on the data front is the development of O*NET by the Employment and Training Administration of the Department of Labor. O*NET will offer data on skills, abilities, and credentials of workers in various occupations, as well as specific descriptions of work performed by occupation. This degree of detail should assist the construction of behavioral models of occupational demand, but O*NET does not provide data specifically about science and engineering occupations at the doctoral level.

Administrators from academia and industry discussed their use of forecasts. John Armstrong, who is a retired vice president for science and technology at IBM, expressed concern that forecasts predicting a decline in demand (e.g., for physicists or hardware engineers) receive much less publicity than those predicting shortages. Although attention to shortages can speed adjustment, industrial personnel forecasting is typically short term and dictated by the annual budget cycle. When a longer-term shortage is anticipated (for example, of polymer scientists in the mid-1980s), industry does not wait for shifts in government policy. Rather, large firms directly fund university programs designed to increase the supply of a particular type of scientist or engineer. There is a need, however, for publicly funded programs to facilitate retraining and for additional research on occupational choice by college students. Because private firms cannot capture all the productivity gain from worker retraining, society will tend to invest too little in this, at some social cost. Both retraining and modifying to initial training that facilitate retraining are areas where there is a social interest that goes beyond the interests of individual firms or scientists.

Ronald Ehrenberg and George Walker agreed that as academic administrators they make little use of forecasts of demand and supply for scientists and engineers. Federal research funding is far more important than student demand in determining graduate funding. At selective institutions, as the attractiveness of under-graduate majors waxes and wanes, the quality of undergraduate majors varies, and the more attractive majors are more selective. Aside from the imposition of enrollment ceilings for the most popular fields, no attempt is made on the part of institutions to control quantity. The quality dimension is also important in faculty hiring. Good forecasts would be helpful here since universities do have some flexibility in the timing of new hiring. For example, even in the absence of mandatory retirement, the replacement cycle in faculty hiring posits an increase in retirements in 15 to 20 years.[1] Good forecasts of Ph.D. wages and quality would be helpful at that time, so that new hiring could be spread out over time rather than occur primarily when demand is greatest (and wages are at their highest level). Of course, if all universities maintain a similar

---

[1] The forecasts of the late 1980s did not anticipate the elimination of mandatory retirement.

replacement cycle and anticipate it in a similar way, the forecasts of shortages are unlikely to materialize. Walker also stressed the need for graduate students to master strong and transferable skills in science and mathematics and to be encouraged to take risks. Since long-term employment prospects are not likely to be predictable, students need to keep their options open. Narrow graduate training is counterproductive in this regard.

Another user with a practical need for long-term forecasts is TIAA/CREF and other retirement plans. Good forecasts could help such organizations allocate their internal resources to reflect the changing mix of customers (e.g., retirees, young faculty). Such forecasts would also better permit these organizations to prepare and target educational materials for the increasingly diverse groups that make up employees in nonprofit education and research institutions (e.g., part-time and adjunct faculty as well as the traditional base of university faculty and staff). For such organizations, the need for sensible projections based on timely data is really a bottom-line issue, not just a question of national or individual decision making. The timing and magnitude of hiring and retirement have implications for cash flow and the term structure of their investments.

The Bureau of Labor Statistics has been producing an "occupational outlook" for many years and Neal Rosenthal discussed this effort. The outlook makes projections for 10 years and is widely used for career guidance by high schools and post-secondary institutions. The general projection technique is described in the box that follows.

The BLS evaluates its outlook five and 10 years after publication and analyzes what went wrong.[2] Over the years it has altered its methodology in the light of these findings, although there are still persistent methodological problems. For example,

---

[2] See for example, *Monthly Labor Review*, November 1997, pp. x-y.

The BLS Occupational Outlooks

The Bureau of Labor Statistics Occupational Outlooks are based on a series of steps that take the forecaster from projections of demand for final goods to projections of demand for occupational employment. In brief, the forecasts are described below.

1. The forecasts begin with the major components of gross domestic product based on commercial macroeconomic models. These models are made for a moderate number of major categories, the spending for which is then attributed to each of a much larger number of commodities.
2. The demand for these commodities is then linked to final demand in each of a large number of industries using input-output tables.
3. Employment by each industry is then calculated by extrapolating the historical trend in the relation of industry person-hours employed to industry output. Person-hours are then converted to jobs by assuming a constant average hours per week in the industry.
4. Employment by industry is converted to employment by occupation using a matrix showing employment in 513 detailed occupations in each of 260 detailed industries. This matrix has the characteristics, although not the same content, of an input-output matrix.
5. Narrow demographic groups forecast labor supply as projections of the labor force based on past trends of participation.

---

there is an implicit assumption that some relationships are unchanging over time. Thus a fixed relationship is assumed in each industry between the number of jobs and total person-hours. This is clearly problematic, since very strong evidence exists that this relationship has changed and will continue to change as the fixed costs of employment rise relative to variable costs and as the relative importance of overtime cost declines. Also worrisome is the assumption that relationships change at the same rate (linear or exponential) as they have in the past. Finally, the BLS outlook neglects many dimensions in which adjustment may occur, including training and retraining, and especially in response to changes in wages. None of the past changes in the relationships is assumed to have been affected by anything behavioral—everything is summarized in the time trend. This does not invalidate the BLS framework as a source of information on structural factors that are likely to drive the future market. However, the limitations of the BLS approach need to be communicated to help users understand that behavioral adjustments have not been included. No response is built into time trends in relative occupational wages on either the demand side (where employers substitute capital for labor when relative wages rise) or the supply side (where students move toward occupations in which relative wages are rising).

Although the BLS techniques can be criticized on methodological grounds, they do provide comprehensive occupational forecasts that are in the public domain, although not for doctoral scientists and engineers. On the other hand, both employers and students can respond to the BLS forecast, making it less likely that the predictions will materialize. Moreover, the omission of behavioral responses makes the BLS outlook unreliable as a basis for decisions on federal funding designed to respond to anticipated shortages.

Another user group is the U.S. Congress. It does not need forecasts on a regular basis, but does need them when an issue arises like the adequacy of the supply of information technology workers. In such situations, forecasts are often produced by groups with a vested interest in the outcome of legislation and a limited technical understanding of the rigors of sampling, forecasting models, and labor market definitions.[3] Even if it is difficult to construct supply and demand models in a legislative time frame, a cadre of analysts who have studied the market and can critique the forecasts of special interest groups is valuable. It would be helpful

---

[3] In the information technology case, an industry group produced projections based on vacancy rates (ITAA, 1998). In an industry that experiences normally high rates of labor turnover, vacancy rates will be high and will overstate the extent of shortages for the industry.

if the conclusions made by forecasting models were presented with all of the uncertainties up front. Politicians will pick and choose the results that serve their desired objectives, but at least they will know the caveats. As Skip Stiles, a staff member for the House Science Committee put it, "There needs to be some way of confusing the issues with the facts." Congress would not only like better forecasts, it would like ways of evaluating the national need for Ph.D. production in science and engineering that is fed by federal funding for research and education. Some of the increases in Ph.D. production recently have come from relatively new doctoral programs. How does that relate to how well science and engineering are being done and implicitly how well taxpayer dollars are being spent? More generally, a better understanding of the higher education system and the incentives that drive it are needed. Modeling should anticipate problems, not just explain past events. Some current trends that are not accounted for in the current generation of supply and demand models are globalization of research, distance learning, education by industry that bypasses traditional institutions of higher education, and the growth of industry/university partnerships.

CHAPTER 4

NEGLECTED MARGINS
OF ADJUSTMENT:
SUBSTITUTION AND QUALITY

Key questions for understanding markets for doctoral scientists and engineers concern the possibilities for substitution and how long it takes for substitution to occur in the presence of shortages or surpluses. Substitution is the process by which supply and demand adjust. Employers may revise job descriptions, reassign tasks, reorganize laboratory operations, and modify training programs. Employees may switch activities or fields and retrain to qualify in new areas. For example, when Ph.D. engineers are plentiful, they may be shifted to administration or sales or be asked to work with less equipment or technical support. When they are scarce, technicians may be asked to take on some of their tasks, and may receive additional training to do this work. Substitution occurs in response to shortages or surpluses or to changes in wages that result from the process of bidding away jobs or workers when the two are not in balance. When substitution occurs slowly, another margin of adjustment will be worker quality. Facing a shortage, employers may accept less talented or qualified employees. If the quality of scientists declines when substitution occurs, the probability of scientific breakthroughs may decrease and development time, errors that require correction and product recall, or the need for quantity of managerial oversight may increase.

Substitution is important on both the supply and demand sides of markets for doctoral scientists and engineers. On the demand side, for example, if employers find it difficult to hire qualified Ph.D.s and the wages of Ph.D.s rise relative to the cost of software, employers may substitute capital or software for labor.

They can also move research operations overseas or encourage foreign scientists to immigrate if enough U.S. scientists cannot be found. On the supply side, students can change fields. Employed scientists with a fundamental skill like mathematics can use that skill working for employers of different descriptions (e.g., in recent years, physicists who were unable to find appropriate jobs in physics have found jobs in the financial services industry).

An understanding of the time period over which substitution can occur is important to the formation of policy designed to encourage or discourage production of new doctoral scientists and engineers. If undergraduates change majors rapidly in order to move into the hottest fields, and if doctoral students can easily change their area of research concentration when new areas emerge, shortages are likely to be short lived. Conversely, if students lack information about what fields are opening up, if significant "retooling" to change fields is required, or if institutional factors inhibit changes of wages, substitution may take a long time to occur, and the result will be a relatively inelastic, short-run, labor supply curve.

The workshop paper by Sherwin Rosen and Jaewoo Ryoo addressed some of these issues in the engineering labor market. Their model consisted of four equations:

1. A demand equation presented as an inverse function, in which the wage an engineering graduate receives depends on the stock of engineers at graduation (the end of the production period—four years for a B.S. in Engineering) and on variables that reflect shifts in demand.

2. A supply equation for new entrants to engineering, in which the number of new entrants depends on the discounted present value of future wages that entrants expect, on variables that reflect the relative attractiveness of other professions, and on the number of new entrants a year earlier.

3. A demographic identity equation, in which the current stock of workers depends on the stock of workers a year earlier less attrition, plus new entrants in the current year.

4. An expected career-prospects equation, in which entry decisions depend on discounted expected future earnings in engineering.

In solving the model, the authors (Ryoo and Rosen, 1998) found that:

> [E]ntry into school of any cohort is negatively related to the stock of practitioners they expect to encounter upon entry at graduation. For example, if many students are currently enrolled in engineering schools, then entry of freshmen in the current period is deterred, *ceteris paribus*. Of course enrollments are encouraged by greater expected future demand conditions and discouraged by greater expected career prospects in alternative occupations.

It should be noted, however, that changes in direction are usually much harder to forecast than longer-term trends. The relative usefulness of such an expectations-based model will depend on the relative accuracy of expectations.

Using data from the *Current Population Survey*, Engineering Manpower Commission, and NSF,[1] Ryoo and Rosen estimated the parameters for the system of equations in the model after normalizing the variables (e.g., the number of engineering graduates relative to college graduates, investment in R&D per unit of GDP, and so forth). They found that supply is very responsive to changes in

---

[1] For more detail on data sources, see J. Ryoo and S. Rosen, "The Market for Engineers," National Bureau for Economic Research, 1992.

demand. Demand is considerably less responsive to changes in R&D expenditures relative to GDP. The authors tried different ways of modeling expectations. Prospective entrants can base their expectations on future wages (assuming that actual future values are the same as estimated future values), on past wages, on both past and estimated future wages, and only on current wages (leading to cobweb behavior). The authors concluded that expectations appear to be forward looking, and this implies that supply behavior will be less volatile than a simple myopic (cobweb) model might suggest.

In the discussion that followed, the Ryoo-Rosen model was commended for explicitly modeling wages and expectations about wages. There may be greater problems, though, in applying such a model to markets for Ph.D.s. Lags are much longer, and there are opportunities for substitution not just between engineering and other professions, but among those who have earned different levels of degrees. Cycles are long, and what may matter most to prospective enrollees are changes in demand. When will demand begin to decline? Lags in the data are such that these turning points are identified in the data only years after they have occurred. A better, but probably unachievable, goal would be to construct models that would predict turning points with some accuracy. Such findings could then be publicized and affect student decisions. In this dimension, science and engineering students may not be much different from majors in art, theater, music, and literature.

Other responses to the model noted that immigration is an important alternative supply source for engineering markets, but one that does not appear explicitly in the model. Further, as Rosen pointed out, much of the demand for engineers comes from the industrial sector and the bulk of demand is for people with baccalaureate degrees. Demand for doctoral scientists is generated to a much greater degree by the educational sector. In the current system, research funding generates a demand for both Ph.D.s and for doctoral and postdoctoral students to staff research projects. In the life sciences, in particular, there may be a long lag between the career decision and undertaking the career itself. Forecasting lifetime earnings, or even earnings seven years into a career, may be far more problematic in these fields than in the market for baccalaureate engineers. An increase in federal spending for grants may create a demand for graduate students that is divorced from market demand for Ph.D.s.[2]

The effect of wage expectations on the supply of Ph.D.s should also consider the uncertainty of the career wage profile, including uncertainties about an individual's ability to attain breakthroughs (discoveries, patents, prizes, tenure, etc.). Any systematic misperception of these probabilities by students, owing to overestimation of their ability relative to others, may lead to unrealistic supply decisions even in the face of evidence that median wages will be low.

The role postdoctoral study plays in the adjustment of the Ph.D. labor market also needs to be examined. There are substantial differences by field in the utilization of postdoctorals. For example, there are very few postdoctorals in the social sciences, whereas in the biological sciences, more than 80 percent of new Ph.D.s go on to postdoctoral study. A period of postdoctoral study postpones entry into the "regular" labor market, and evidence exists that the length of this delay is increasing (NRC, 1998). For new Ph.D.s, a postdoctoral appointment delays entry into the labor market, thus disrupting supply and shortening the length of careers in regular employment. Also, the postdoctoral pool provides a ready supply of highly trained Ph.D.s that is available for regular employment if academic or industrial demand increases, thus shortening the period of adjustment until such time as the postdoctoral pool is drained. Models of supply that take into account only enrollments and graduates will miss an important margin of adjustment if they ignore postdoctorals in those fields where the numbers of these positions are large or growing.

---

[2] This point is also made in the Massey and Goldman (1995) paper.

Finally, it was noted that possibilities for substitution may vary by employment sector. Scientists and engineers in industry may be assigned to a variety of projects that blur distinctions based on field. By contrast, in academia a faculty member rarely changes departments and is hardly ever hired in a field outside the field of training.

PRESENTATION OF UNCERTAINTY AND USE OF FORECASTS WITH EXPLICIT UNCERTAINTY

All models are wrong; some models are useful.

George Box

Any forecasting effort should communicate two things: the usefulness of the model and its limitations. How to present uncertainty is essentially a question of how to present the limitations of a model. A forecast must depend on assumptions about the future values of variables that drive the model. How certain are we about those assumptions? For short-range forecasts, we may be quite certain that the expected values of the variables will materialize. As we peer further into the future, such certainty declines. Thus, short-term and long-term forecasts have very different properties, very different uses, and very different levels of uncertainty. A short-term forecast, a one- or two-year forecast using annual data, can do a good job of forecasting point estimates, such as the expected number of engineering graduates in the year 2000. These future graduates are already sitting in classrooms and rates of attrition are relatively constant over time.

Increasing the span—the level of aggregation over occupations—typically increases forecast accuracy because random errors and movements between the occupations in the aggregate are averaged out. However, this may not be the case if increasing span increases heterogeneity, and the forecasting model is not successful in representing the effects of this heterogeneity.

Nancy Kirkendall of the Office of Management and Budget discussed different approaches to presenting the uncertainty of forecasts that are used in the federal government. Different measures of error are appropriate to forecasts with different time frames. For short-term models, the uncertainty of point estimates can be measured by the mean squared error, in which forecasted values are compared with actual observations. Other measures of uncertainty include the confidence interval and mean absolute percentage error. These simple methods do a good job of representing uncertainty. Some forecast competitions have shown that these are the best methods to measure uncertainty for short-term forecasts.

Longer-term forecasts are more complex than short-term forecasts and are more vulnerable to unanticipated changes in the economic environment. Forecasts made for a longer time span are typically less accurate than those made for a shorter span.[1] Although the mean squared error may be computed for long-term forecasts, it may not be a good measure of the model's usefulness. The Bureau of Labor Statistics (BLS) routinely assesses the accuracy of its forecasts. The example below shows the systematic reduction in the forecast's percent error as the time span of the forecast decreases. It is possible to take these errors and attribute them to factors unknown at the time of the forecast, such as immigration rates.

---

[1] There may be exceptions to the relative accuracy of long- and short-term forecasts due to the volatility in conditioning variables. For example, annual rainfall may be forecasted with better percentage accuracy than rainfall on any particular day because short-run volatility tends to average out over the long run.

Example

The following tabulation shows BLS forecasts for the civilian noninstitutionalized population aged 16 and over for the total population and the errors associated with the total population forecast:

Forecast of 1995 Population Made in	Number of People (in Millions)	Percent Error
1980	186	−6.3
1983	194	−2.4
1985	194	−2.4
1987	196	−1.4
1989	196	−1.4
1991	196	−0.4
1995	199	• • •

The Energy Information Administration (EIA) publishes a short-term forecasting system that projects energy prices for one to two years. The EIA computes the mean squared errors for these different forecasts and uses the information to fine-tune its models. However, for long-term forecasts they do not publish the error terms. Because the accuracy of the forecasts would not be very good, long-term forecasts should not be used to produce point estimates. Long-term forecasting models are useful for other purposes. The examples that follow demonstrate some of these uses.

Scenario analyses are useful in illustrating the variability and uncertainty of long-term forecasts. An example is the case of natural gas forecasting. Figure 5-1 shows the historical trend in natural gas wellhead prices and forecasts future prices with a reference case and two additional scenarios. The scenarios reflect changes that might happen in the forecast depending on the assumptions made on how rapidly new technology penetrates the

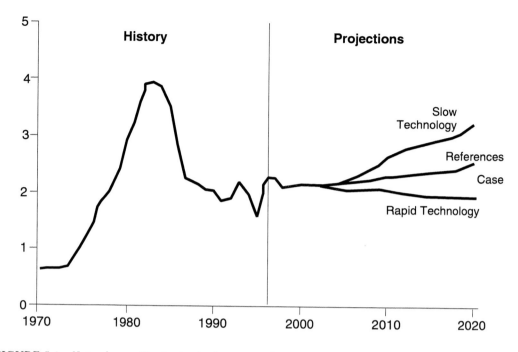

FIGURE 5-1 *Natural gas wellhead prices in three cases, 1970–2020 (1996 dollars per thousand cubic feet).*

market. The scenarios include high economic growth (2.5 percent annually) and low economic growth (1.3 percent annually). The reference case posits 1.7 percent growth per year. Scenario analysis can be quite instructive if past data are highly variable. Past variability communicates to the user that the assumption of smooth patterns of change is unwarranted but that volatility may not be predictable. Sometimes scenarios are constructed from statistical confidence bounds on model inputs or parameters, such as confidence limits on the future economic growth rate. Users can be misled, however, if they interpret the resulting scenarios as confidence bounds, as neither future paths nor forecasts at particular dates need be contained between "high" and "low" scenarios with any degree of confidence. This is particularly an issue when the "high" and "low" scenarios are a concatenation of events that are not necessarily independent of one another. Finally, the scenarios can illustrate a range of outcomes under the assumption that there are no large shocks.

Providing a model in which users can adjust inputs is another way to help users understand the effects of uncertainty in forecasts. For example, in 1997 the EIA made a series of assumptions in its forecasts of the price of electricity. The *Annual Energy Outlook 1997* addressed electricity restructuring by incorporating the Federal Energy Regulatory Commission actions on open access, lower costs for natural gas-fired generation plants, and early retirements of higher-cost fossil plants. *The Annual Energy Outlook 1998* makes additional assumptions about competitive pricing and restructuring, including:

- Lower operation and maintenance costs.
- Lower capital costs and improved efficiency for coal- and gas-fired plants.
- Lower general and administrative costs.
- Early retirement of high-cost nuclear plants.
- Capital recovery period will be reduced from 30 to 20 years.
- California, New York, and New England will begin competitive pricing in 1998 with stranded cost recovery phased out by 2008.

As illustrated in Figure 5-2, because of the additions to and changes in model assumptions and lower projected coal prices, average forecasted electricity prices for 2015 are 13 percent lower than were forecasted in 1997.

There are four fundamental sources of uncertainty in the forecasts made for the availability of scientists and engineers. The first may be identified as exogenous variables, factors that are outside the labor market for scientists and engineers but never-

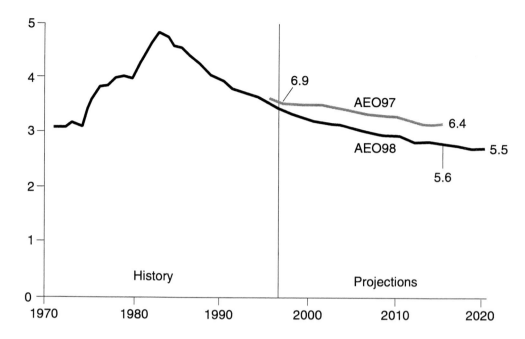

FIGURE 5-2 *Electricity prices, AEO98 vs. AEO97, 1970–2020 (1996 cents per kilowatt-hour).*

theless affect it. These include economic growth both in the United States and elsewhere in the world, technology growth, defense needs, wars, and the demographics of scientists and engineers. We are reasonably good at describing the overall demographics of the population, but scientists and engineers are a special group that is affected by things that we are not good at predicting, such as labor force participation and immigration. The demographics of scientists and engineers may also be affected by the changing ethnic and gender composition of the work force. To date, white and Asian American males have been far more likely to go into careers in science and engineering than members of other groups, which make up an increasing share of the student population. Finally, we need to be concerned about the number of people who have sufficient preparation in mathematics to enter the scientific and engineering fields.

A second source of uncertainty stems from factors that are subject to policy control but that do not necessarily accord with the interests of optimizing the labor market for scientists and engineers. For example, to some extent, we know that immigration policy incorporates concerns about the market for scientists and engineers. However, immigration policy is also directed toward a number of other objectives. Moreover, government subsidies may also affect the market for scientists and engineers. In addition to the funding for research and education provided by NSF and NIH, spending on defense contracts, defense-related research, and funding of medical schools through the Medicare program must be considered. While these variables are somewhat subject to policy control, they are directed toward goals other than ensuring that the market for scientists and engineers remains stable or grows at some desired rate.

The third source of uncertainty is behavioral uncertainty, which comes from our inability to predict perfectly how people will respond to the market. These areas of uncertainty include attitudes of college students toward science, plans of the scientists or engineers for shifting from scientific work to other tasks such as administration, and attitudes toward retirement.

The fourth source of uncertainty is the most serious and the hardest to convey to forecast users—what economists call parameter uncertainty (also called systematic errors), or uncertainty in the estimated model itself. This uncertainty relates to both our inability to capture all the nuances of the real world in our models (parameter uncertainty) and limits our ability to calibrate our models perfectly with limited data (model uncertainty). We would like to believe that we could build a structural model that really incorporates all of the behavioral decisions that people make when they choose to enter science or engineering. In fact, we have neither the data nor the behavioral laws to permit the construction of such a model. Our models are not true structural models, and their parameters change as adjustment occurs at neglected margins.

For example, in recent years when the wages for scientists, engineers, or other technically trained people rose, employers divided jobs into components. Some of these components did not require a highly trained analytical person and could be given to people who had lesser or more simpler technical training. Variation in the skill requirements needed to achieve a particular kind of production is not typically incorporated in the models we build. In addition, it is difficult to measure the potential of using technology to make one person able to do a job previously done by two people. These measures of quality and substitution are very difficult to quantify.

Forecasts should be designed for a specific objective. We recognize that the labor market for scientists and engineers is not a classic spot market in which workers offer their labor in response to a wage offer by employers and higher wage offers immediately bring forth a supply of additional workers. Were it so, the demand for labor by government and industry would be quickly met by a supply of scientists and engineers and the market would clear easily. In actuality, the NSF, NIH, and others who really need these forecasts the most are making long-term decisions. They are trying to decide how many graduate fellowships and traineeships to provide in order to create opportunities for people who years later may become scientists and engineers.

Targeting quantity with sufficient lead time for consumers of these forecasts is very difficult. However, it is less difficult to assess whether the wages of scientists and engineers are similar to those of other personnel with comparable aptitude (based on SATs and ACTs) who take jobs where there is some substitution with science and engineering. It is not difficult to know whether the salaries of scientists have stayed constant while the salaries of other groups with similar years of training (for example, lawyers) have risen (see Figure 5-3).

We should also examine possible sources of short-run adjustment. Institutions could make the market for scientists and engineers more like a spot market. These actions include:

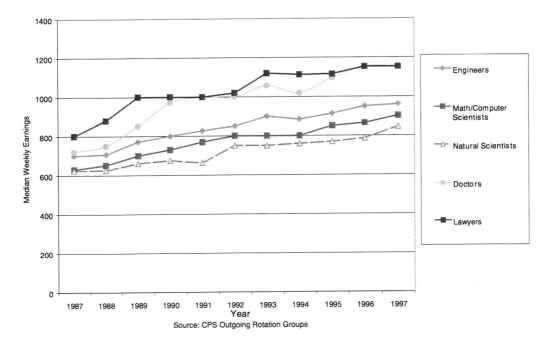

Source: CPS Outgoing Rotation Groups

FIGURE 5-3 *Median weekly earnings for selected occupations by year.*

(1) development of more flexible programs for the use of "temporary" immigrant scientists and engineers; (2) development of more stable and humane versions of the postdoctoral pool that would make these individuals more productive in research and prepare them to move in and out of employment in specific specializations depending on market conditions; and (3) institutional ceilings on enrollments that are responsive to short-term market forecasts. Usually this final measure is destabilizing, since institutions have difficulty expanding or contracting faculty size in the short term.

A distinction should be made between the instruments of policy control (e.g., enrollment quotas) and the targets of policy control, e.g., the number of research personnel. While the former are subject to rapid change, the latter are not likely to be characterized by volatility or disruptions that have extreme private or social costs. One not only needs to understand how policy affects behavior, but also to project how short-term behavioral adjustments translate into long-term market conditions.

Forecasters need ways to reassure the users of their forecasts that what they are doing is acceptable, correct, and scientific. They need to describe forecasts more clearly and to describe how they should and should not be used. Finally, forecasters should document exactly what they do. Even with very clear documentation, some skeptics will assume that the forecasters have fiddled with the numbers.

A concomitant issue develops from users of forecasts for scientists and engineers. Some users may ignore uncertainty and convert an estimate that is presented as uncertain into a point estimate (Kahneman et al., 1982). Moreover, they may subjectively or unconsciously amplify the probability of rare events. Finally, they may tend to overemphasize losses as opposed to gains. These behaviors may explain why forecasts are often considered by users to be more certain than forecasters intended and why forecasts of a shortage or glut of scientists and engineers receive far more public attention than justified by their estimated probabilities.

The confusion of users about how to use forecasts argues for conducting research to determine the best ways to present forecasts. Several statistical agencies have started cognitive laboratories that evaluate questionnaires. They examine the wording of questions to find out whether people understand what they are being asked and whether they give accurate answers. This helps agencies to fine-tune the questions and tailor the responses to address agency goals. Perhaps the forecasting community could conduct cognitive studies on how best to present a forecast's usefulness and accuracy. Research of this type could provide good scientific information about what does and does not work in terms of presenting forecasts.

In summary, to present uncertainty in ways that will not mystify users, several steps should be taken:

- First, the user should know if the forecast is conditional or unconditional—whether the forecaster is assuming that the future values of the explanatory variables are known with certainty (unconditional) or whether these values are uncertain, so that the forecast is conditional on whether these values materialize.[2]

- Second, the forecaster needs to consider exactly what the model is expected to accomplish. Is the forecast short term or long term? If the model is short-term, supply is constrained by what is already in the pipeline and demand is substantially constrained by commitments to budgets and research programs. Long-term market forecasts are made for time periods where there are substantial problems in forecasting "fundamentals," such as choice of college major and national research priorities and funding. With long-term forecasts the focus should not be on actual forecast accuracy—although accuracy within some limits is desirable.

- Third, forecasting needs to be a continuous, dynamic process in which past performance is tracked, and the forecasters learn from their past mistakes.

- Fourth, the forecaster must decide on the objectives of the forecast. Should the forecaster conduct a scenario analysis? To what factors should the forecast be sensitive?

- Fifth, the forecaster must ensure that the required data are available.

- Sixth, once committed to the forecast, the forecaster must quantitatively evaluate the model against the objective function in order to assess the model's accuracy. If the model is inaccurate, it should be respecified.

---

[2] An example of an almost unconditional forecast is the number of 18-21 year olds six years from now, since we are almost certain about age-specific rates of mortality and we know how many 12-15 year olds there are now. A forecast of college enrollment for 18-21 year olds, however, is likely to be conditional, since it depends on rates of high school completion and college participation that are uncertain and may depend on variables outside the pure demographic model, e.g., job prospects for high school graduates and availability of financial aid.

- Seventh, the data and the methodology for the model must be well documented. The documentation needs to be widely available and readable, and all the test results should be made available to all users.

- Eighth, cognitive studies should be used to provide input on how best to present the forecasts and the information. Employ policymakers and staff to participate in a cognitive study to help refine information and forecast presentation.

- Finally, forecasters need to conduct research into methods of displaying uncertainty of forecasts. The goal is to discover the most convincing ways to display forecast results so the audience understands that forecasting models are a useful tool in the policy formation process, but have limitations in accuracy that affect how they should be interpreted and used.

CHAPTER 6

SUMMARY AND RECOMMENDATIONS

Summary

The committee was charged to make recommendations on the government's optimal role in forecasting the supply and demand of scientists and engineers, and in particular whether NSF itself should be involved in forecasting and related activities such as data collection. Throughout the workshop, speakers, discussants, and participants addressed a number of salient issues. These issues are synthesized here in the form of five questions followed by the consensus responses that developed throughout the workshop.

1. Who are the clients for forecasts of demand and supply of doctoral scientists and engineers, and what is it that they really need and want? Two primary groups have been identified: (1) students who are deciding on careers and (2) funding agencies, such as NSF and NIH, that must decide how to allocate funds for traineeships and research assistantships. Other clients include universities that face decisions on the size of research and teaching programs and faculty recruitment and members of Congress who make policy decisions that affect research and the labor market. These other clients need data and forecasts that are subsets of those needed by NSF and NIH.

2. What is the appropriate scope for forecasts? Of course the answer will be different depending on who the client is, but six questions that help define the scope have been identified.

- First, should the forecasts be conditional or unconditional? Unconditional means forecasting what is going to happen in the future, all things taken together. Conditional means a what-if analysis; if NIH increased funding for some areas of biology, how would that impact the market? Both kinds of forecasts may be needed, but the distinction must be made.

- Second, at what level should forecasts be disaggregated? There has been discussion about the need for very disaggregated forecasts, such as the demand for people in informatics. There have been other discussions about the impossibility of disaggregation when there is so much cross-field substitutability that only at an aggregate level does the forecast make sense.

- Third, what variables should be part of a forecast? The NSF forecasts that have been criticized for concentrating only on numbers of jobs or the supply of job seekers. Other important variables should be considered (1) salaries, which economists identify as the most important element in the operation of markets; (2) the quality of the people that are going into these markets; and (3) the nature of the work that they do, whether it is in bench science, administration, marketing, or whatever.

- Fourth, how much detail is needed to track career paths and labor market changes by sector of employment? The greatest attention has been concentrated on the academic market, obviously a very important market for scientists and engineers. However, the industrial market is a large and growing market. The level of detail required in forecasts of industrial employment of scientists and engineers has not been resolved. For example, is it useful to produce separate forecasts of demand for biologists in the agricultural and pharmaceutical industries, or does a forecast of overall industrial demand suffice?

- Fifth, what is to be described? Are you forecasting a series of snapshots of the market in the future? Are you trying to describe the flows and the transitions, or are you trying to tell a story about future career expectations for scientists and engineers?

Those different goals will have different implications in the establishment of models.

•   Sixth, how should uncertainty be modeled and presented? Many of the variables we have discussed are distributions, such as the distribution of salaries. Making a point forecast of mean salary may not meet all the clients' needs. The dispersion of salaries may need to be modeled. How is information about distributions to be presented, and what is the best way to present information about the inherent uncertainty in the models?

3. Complaints about the quality of the data that are available were a recurring theme during the workshop. What data do we actually have now, and what are their deficiencies? What can be done to improve the utility of the data that we are currently collecting? In addition, what new data are needed? What are the critical gaps in the current data collection programs? Are data access and data documentation adequate? What major steps must be taken to make the data that we currently collect more useful? Finally, what about data management and coordination between different collectors of data? What needs to be changed there?

4. The fourth question relates to modeling issues. To some extent, the variables included, the details of the model, and what is being forecasted will determine much about the model itself, but there are still additional modeling issues.

•   What are the drivers, those things that are outside of science and engineering of which we need to take account? Supply demographics and research and development funding have been emphasized in this regard.

•   What variables should be included? Economists talk about prices, salaries, jobs, and the quality of the people in the market. In addition, should job vacancies or underemployment be considered? Economists view these phenomena as temporary symptoms of market disequilibrium that market forces are always pushing to eliminate rather than "permanent" characteristics of markets.

• What should the nature of the model be? Gap models that exclude equilibrating mechanisms (through salary adjustments, changes in job description and tasks, acceleration of research programs, and so forth) and orthodox economic models that (in principle) include these equilibrating mechanisms are perceived as more different than they actually are. Gap models identify the size of the problem that equilibrating mechanisms must handle but often fail to specify which equilibrating mechanisms are operating and how quickly they will work. Users of gap models sometimes make the *de facto* assumption that equilibrating mechanisms are very slow. Orthodox models often assume that equilibration is rapid without articulating the specific mechanisms that produce equilibrating adjustments or determining that they do indeed work quickly. The resolution of the differences between these modeling approaches requires careful identification and empirical analysis of equilibrating mechanisms and their speed of operation.

• Regarding model complexity, there is a tendency in all of science and engineering to make things more complex on the grounds that complexity is necessary for realism or accuracy. Certainly forecasting models have become very complex. Considering whether that is the best approach may be useful, especially since forecast users are not economists. Openness in models with some transparency in their structure that allows users to access those models at various levels is also desirable. Having a more realistic but highly complex model may work at cross-purposes with client needs.

• A good model of this market clearly needs to draw on behavior. People respond to incentives in this market, and we need to understand the quantitative nature of those responses. There are many ways to do this. Economists traditionally prepare a mathematical, econometric model and use historical data to infer what the behavioral response parameters might be. Another approach, more like a classical scientific field experiment, examines the impact of an increase in the number of traineeships on Ph.D. completion rates, the behavior of postdoctorals, and so forth.

5. Finally, what is the best administrative structure for forecasting? A fundamental question is whether the government should be making official forecasts in this market. In particular, should NSF and its Science Resources Studies (SRS) group be in the business of making an official forecast? Does an official forecast do more harm than good in encouraging market adjustment? If there is an official forecast, how should it be structured to protect NSF from either the fact or the appearance of political interference? On a related issue, if there is no official forecast, what role should the SRS section of NSF and the government play in forecasting? Data collection on scientific and engineering personnel has been in the governmental domain and presumably will remain there. Should a clear division be made between the agencies that collect data on this market and agencies that are engaged in forecasting or policy analysis? Could this division then facilitate the independence of the agencies involved or maintain the integrity of the data collection efforts? In one model, for example, NSF would not make an official forecast, but similar to blue-chip indicators would present a variety of forecasts prepared by others.

Recommendations

The forecasting process is ongoing. Forecasters must learn from their mistakes. The whole forecasting exercise needs to be placed within an administrative framework that facilitates an evaluation process and a process to correct errors. The Science Resources Study Division (SRS) of the NSF is the locus of action in the federal government to bring about the improvement of data and forecasts of the market for doctoral scientists and engineers. The committee has therefore directed most of its recommendations to SRS, since this division should be able to encourage improvements in the construction and use of forecasting models for highly trained scientists and engineers, even when SRS should not carry out the work itself.

Recommendation 1. The producers of forecasts should take into account the variety of consumers of forecasts of demand and supply for scientists and engineers.

NSF should recognize that there are five distinct communities of clients for data and projections on the supply and demand of scientists and engineers, each with different needs and interests. NSF's data collection and forecasting activities should keep the needs of these different communities in mind. They are:

1. students making career decisions,
2. federal, state, and private funding agencies,
3. industrial and academic employers of scientists,
4. Congress in its role as funder and policymaker, and
5. the scientific community that conducts research and produces studies of the market and its participants.

Representatives of all these user communities spoke at the workshop. All expressed dissatisfaction with the current state of data and forecasting.

Students need qualitative projections of likely career outcomes and probabilities of success, with particular attention to the state of the job market in a few years when they will seek employment. These qualitative projections require timely data, but they need not be based on broad surveys or censuses.

Research funding agencies and Congress need relatively long-term projections of supply and demand factors by specific discipline that can be used to guide policy on training support and institutional development. These projections should take careful account of the ease with which one kind of labor can be substituted for another and the incentives and behavioral responses that operate. The importance of contingencies and of forecast uncertainty should be emphasized.

The needs of employers are varied, but they also benefit from an early warning about shortages. As producers of Ph.D.s, academic institutions can take steps to expand or contract Ph.D. enrollment given convincing evidence of emerging labor market trends.

Finally, the forecasting community should be more forthcoming about the appropriate use of forecasts and the nature of their underlying assumptions.

Recommendation 2. The NSF should not produce or sponsor "official" forecasts of supply and demand of scientists and engineers, but should support scholarship to improve the quality of underlying data and methodology.

NSF should not produce or sponsor "official" forecasts of supply and demand in the markets for scientists and engineers, but should continue to take the lead in collecting and making available data on these markets. A clear organizational separation should be made between data collection and modeling/forecasting activities undertaken for NSF's own policy use or for use by federal agencies. For example, convert the SRS into a National Center for Science Statistics on the model of the National Center for Health Statistics (NCHS), the National Center for Educational Statistics (NCES), or the Energy Information Agency (EIA) and remove modeling and forecasting activities to a separate policy unit. Or, for example, forecasts could be produced by an outside agency with statistical expertise. Agencies such as the Bureau of the Census or the Bureau of Economic Analysis may be well suited to undertake such forecasts.

If asked to produce forecasts of scientific and engineering personnel for its own use or the use of other agencies, the NSF policy unit should avoid endorsing or emphasizing "gap" models that do not incorporate behavioral adjustment to demand and supply and consequently may give unwary users a misleading impression of likely market outcomes. NSF should avoid suggesting that there is a single best level of detail and model complexity for the forecasts needed by various users and should instead maintain that model structure will depend on user needs and objectives.

The committee reviewed the history of NSF projections, most notably those in *The State of Academic Science and Engineering*. In response to the perception that there could be a connection between NSF funding and its projections, especially regarding projections of shortages, the committee believed that NSF should limit itself to data collection and dissemination and use external "arms-length" forecasts for its policy needs to avoid the possible conflict of interest that might occur if it produced its own forecasts.

Recommendation 3: Undertake a comprehensive review of data collection in the light of forecasting needs.

NSF's SRS should undertake a comprehensive review of its data management program, preferably in coordination with Bureau of Labor Statistics (BLS), NCHS, and NCES. It should be to seek the production of more timely and useful data on the market for scientists and engineers. In addition, it should coordinate definitions and categories across agencies to facilitate a consistent picture of the different stages in the market, from student training and degree choice to mid-career transitions across and out of science and engineering fields. Moreover, sample sizes have been reduced since the late 1970s, which makes modeling difficult for small fields, specific employment sectors, and for rare events (such as mid-career changes).

Recommendation 4. Data that enhance forecasts should be widely available and be disseminated on a timely basis.

NSF's SRS should establish three high-priority data objectives. These are: (1) production of timely, descriptive statistics on employment and salaries by field of training, occupation, and sector, in a consistent time-series format that permits tracking and projections of trends; (2) production of an individual-level Public Use Sample, containing a consistent time series of cross-sections of doctoral recipients, and when available, nondoctoral recipients; and (3) production of a Public Use dataset panel of scientists and engineers to analyze transitions from the educational system to employment and transitions across fields, occupations, and activities. NSF should process these data, establish rules for access so that they are widely available, and institute less burdensome mechanisms to protect confidentiality. For example, in producing a Public Use panel, NSF might recruit respondents who are willing to provide *vitae* in a standard format without confidentiality restrictions. This format could be made available on the seb; to make the process even easier for respondents, NSF could code respondent *vitae*. This would enhance modeling of individual and institutional behavior in response to changes in funding, demographically driven demand, compensation, etc. Standard coding of *vitae* would permit collection of more detailed data than the *Survey of Doctorate Recipients* form.

Both modelers and policy advisors at the workshop complained that data were not timely. At present, there is no way of knowing in a timely way whether market mechanisms are working to alleviate shortages or gluts. Policies calibrated to outdated evidence may provide too little too late, or too much too late. Furthermore, if indicators show that market mechanisms are impacting rapidly, the need for policy change might be obviated altogether.

NSF's/SRS should collect and disseminate data that

enables a variety of forecasting exercises that differ along some or all of the following dimensions:

1. Unconditional vs. conditional ("What if") forecasts.
2. Multiple levels of field disaggregation, e.g., physical sciences, physics, solid state physics, materials science.
3. Optional variables to forecast, e.g., jobs, salaries, quality, productivity, occupation.
4. Various sectors to forecast, e.g., academic tenure track/ other, postdoctoral positions, or industrial sectors.
5. What to forecast, e.g., stocks, flows, transitions, or careers.
6. Permit forecasts to go beyond means and standard errors to more complete descriptions of the distributions of possible outcomes.
7. Various time horizons for forecasts.

Finally, the NSF's SRS Website currently focuses almost exclusively on providing relatively simple tabulations that are useful for casual policy analysis but not very useful for either career planning by students or for research on the science and engineering market. The data management program and website should be redesigned to service these neglected user communities (or in the case of students, the public and private organizations and associations that provide career guidance) and to provide links to other data sources from BLS and NCES that are important for analysis of the markets for scientists and engineers.

Recommendation 5. NSF should develop a research program to improve forecasting.

The Directorate for Social, Behavioral and Economic Sciences of the NSF should commission behavioral studies of scientists and engineers early in their careers, as well as studies of forecasting methods and evaluations of past forecasting exercises.

Particular emphasis should be placed on critical parameters of market response, such as wage-sensitivity of field and occupation switching, and the determinants of the ability of employers to restructure research jobs in response to supply conditions in various fields. Emphasis should be placed on the difficult issues of measuring the quality and productivity of scientists and engineers, the quality of worker-to-job matches, and quality of life for scientists and engineers. This work should be conducted through the ordinary peer-reviewed research support process already in place at NSF. SRS should facilitate the dissemination of results of these studies but should stop short of sponsoring or endorsing specific forecasts or methods.

REFERENCES

Arrow, Kenneth J., and William M. Capron
    1959   "Dynamic shortages and price rises: The engineer-scientist case,"
            Quarterly Journal of Economics (May): 292-308.

Atkinson, Richard C.
    1990   "Supply and demand for scientists and engineers: A National crisis in
            the making," *Science*, 248(April 27):425-432.

Barnow, Bert S.
    1998   "Objectives and Approaches of Forecasting Models for Scientists and
            Engineers." A paper presented at the Workshop on Improving Models
            of Forecasting Demand and Supply for Doctoral Scientists and
            Engineers. National Research Council, Washington, D.C.

Bowen, William G., and Julie Ann Sosa
    1989   *Prospects for Faculty in the Arts and Sciences: A Study of Factors Affecting
            Demand and Supply, 1987 to 2012.* New York: Princeton University
            Press.

Cain, Glen G., Richard B. Freeman, and W. Lee Hansen
    1973   *Labor Market Analysis of Engineers and Technical Workers.* Baltimore:
            Johns Hopkins University Press.

Cartter, Allan M.
    1970   "Scientific manpower for 1970-1985," *Science*, 172(April 9):132-140.
    1976   *Ph.D.s and the Academic Labor Market.* New York: McGraw-Hill.

Ehrenberg, Ronald G.
    1991   Academic labor supply in Charles T Clotfelter, Ronald G. Ehrenberg,
            Malcolm Getz, and John J. Siegfried. *Economic Challenges in Higher
            Education.* Chicago: University of Chicago Press.

Freeman, Richard B.
    1971   *The Market for College-Trained Manpower.* Cambridge, Mass.: Harvard
            University Press.
    1976   "A cobweb model of the supply and starting salary of new engineers,"
            Industrial and Labor Relations Review, 33:236-248.

ITAA (Information Technology Association of America)
1998   *Help Wanted 1998: A Call for Collaborative Action for the New Millennium.* Arlington, VA: ITAA.

Johnson, George
1998   "How Useful are Shortage/Surplus Models of the Labor Market for Scientists and Engineers?" A paper presented at the Workshop on Improving Models of Forecasting Demand and Supply for Doctoral Scientists and Engineers. National Research Council, Washington, D.C.

Kahneman, Daniel, Paul Slovic, and Amos Tversky
1982   *Judgement Under Uncertainty: Heuristics and Basics.* New York: Cambridge University Press.

Kirkendall, Nancy
1998   "All Models Are Wrong; Some Models Are Useful." An overhead presentation made at the Workshop on Improving Models of Forecasting Demand and Supply for Doctoral Scientists and Engineers. National Research Council, Washington, D.C.

Leslie, Larry R. and Ronald L. Oaxaca.
1993   "Scientists and Engineering Supply and Demand" in *Higher Education: Handbook of Theory and Research*, Vol. IX. New York: Agathon Press.

Massy, William F., and Charles A. Goldman
1995   *The Production and Utilization of Science and Engineering Doctorates in the United States*, Stanford Institute for Higher Education Research, August.

NAS (National Academy of Sciences/National Academy of Engineering/ Institute of Medicine)
1993   *Reshaping the Graduate Education of Scientists and Engineers.* Committee and Science, Engineering, and Public Policy. Washington, D.C.: National Academy Press.

NRC (National Research Council)
1994   *Meeting the Nation's Needs for Biomedical and Behavioral Scientists.* Office of Scientific and Engineering Personnel. Washington, D.C.: National Academy Press.
1998   *Trends in the Early Careers of Life Scientists.* Committee on Dimensions, Causes, and Implications of Recent Trends in the Careers of Life Scientists. Washington, D.C.: National Academy Press.

NSF (National Science Foundation)
1989   *Future Scarcities of Scientists and Engineers: Problems and Solutions.* Division of Policy Research and Analysis, National Science Foundation, Washington, D.C.

1989   *The State of Academic Science and Engineering.* NSF 90-35. Washington,
       D.C.: National Science Foundation

Ryoo, Jaewoo and Sherwin Rosen
    1998   "The Engineering Labor Market." A paper presented at the
           Workshop on Improving Models of Forecasting Demand and Supply
           for Doctoral Scientists and Engineers. National Research Council,
           Washington, D.C.

Sovern, Michael I.
    1989   "Higher Education—The Real Crisis." *The New York Times Magazine*,
           January 22, Vol. 137, p. 24.

WORKSHOP
AGENDA

**Workshop on Improving Models of Forecasting
Demand and Supply for Doctoral Scientists and Engineers**

MARCH 19-20, 1998

Lecture Room
National Academy of Sciences
2101 Constitution Avenue, N.W.
Washington, DC

Office of Scientific and Engineering Personnel
National Research Council

MARCH 19, 1998
1:00pm
Welcome

Michael Teitelbaum
Alfred P. Sloan Foundation

Daniel McFadden
University of California, Berkeley

Jeanne Griffith
National Science Foundation

1:30-3:00pm
Panel I
Forecasting Models: Objectives and Approaches

What would the characteristics of good models of supply and demand for scientists and engineers be? Such models would likely recognize that adjustment can occur in three dimensions: quantity, price, and quality. To date, models have focused only on quantity. What models can actually be estimated, however, depend on what data is available, not just rich specification. This session will focus on alternative approaches to models and describe what models could be estimated, given available data. It will also look at what additional data would make possible improvement in the kinds of models used to describe adjustment in these labor markets.

Presenter:
Burt S. Barnow
Johns Hopkins University

Panel:
Ronald Ehrenberg
Cornell University

George Walker
Indiana University

John A. Armstrong
IBM, *retired*

Chair:

Ronald Oaxaca

University of Arizona

3:15-4:45pm
Panel II
Neglected Margins: Substitution and Quality

Much modeling of labor markets for scientists and engineers neglects price, price elasticity, substitution among different types of labor and other variables (corporate restructuring, use of different kinds of academic workers, effects of immigration, etc.). What are the external drivers of substitution? What data might serve as indicators?

Presenter:

Sherwin Rosen

University of Chicago

Panel:

Michael Finn

Oak Ridge Institute for Science and Education

Eric Weinstein

Massachusetts Institute of Technology

Paula Stephan

Georgia State University

Chair:

Daniel Hamermesh

University of Texas, Austin

5:00-7:00pm
Reception

Members Room

MARCH 20, 1998
8:30am
Continental Breakfast

Lecture Anteroom

9:00-10:30am
Panel III
Models of Scientific and Engineering Supply and Demand:
History and Problems

This session will focus on shortage/surplus or "gap" models that have been estimated in the past. Have they been useful? For what purposes? Given the importance of external shocks in these markets, how should the importance of inherent uncertainty be conveyed to non-expert users? Could these models be modified to take into account the simultaneous adjustments of quantity, quality, and price in either equilibrium or adjustment models?

Presenter:
George Johnson
University of Michigan

Panel:
Geoff Davis
Dartmouth College

Charles A. Goldman
RAND

Robert Dauffenbach
University of Oklahoma

Sarah E. Turner
University of Virginia

Chair:
Brett Hammond
TIAA-CREF

10:45-12:30pm
Panel IV
Presentation of Uncertainty
and Use of Forecasts with Explicit Uncertainty

What is the best way to communicate the sensitivity of model outputs to assumptions and to uncertainty? How should uncertainty be presented and explained to policymakers and others who are educated but not expert users? How and why are policymakers likely to misunderstand and "misuse" forecasts and what can be done about it by modelers/forecasters?

Presenter:
Nancy Kirkendall
Office of Management and Budget

Panel:
Daniel Greenberg
Bureau of Labor Statistics

Neal Rosenthal
Science & Government Report
Johns Hopkins University

Chair:

Caroline Hoxby

Harvard University

12:30-1:30pm

Lunch

1:30-3:00pm

Discussion Panel

Should we continue forecasting S&E supply and demand?

Should forecasts be presented differently?

Should we continue forecasting supply and demand for scientists and engineers? Should forecasts be presented differently?

Moderator:

Daniel McFadden

University of California, Berkeley

Participants:

Alexander H. Flax

Institute for Defense Analysis, *retired*

Skip Stiles

Committee on Science

U. S. House of Representatives

Open discussion

BIOGRAPHICAL INFORMATION ON WORKSHOP PARTICIPANTS

John Armstrong is a retired Vice President of Science and Technology at International Business Machines. He has held several positions for IBM, including manager of materials and technology development at the IBM East Fishkill laboratory, and Director of Research. In 1989 he was elected a member of the Corporate Management Board and named Vice President of Science and Technology. Dr. Armstrong's research focused on nuclear resonance, nonlinear optics, the photon statistics of lasers, picosecond pulse measurements, the multiphoton spectroscopy of atoms, the management of research in industry, and issues of science and technology policy. He received an A.B. (summa cum laude) in Physics from Harvard College in 1956 and a Ph.D. from Harvard in 1961.

Burt S. Barnow is Principal Research Scientist at the Institute for Policy Studies at the Johns Hopkins University where he teaches courses in program evaluation and labor economics. Dr. Barnow has 25 years of experience as an economist in the fields of employment and training, labor economics, and program evaluation. Dr. Barnow joined the Institute for Policy Studies in 1992 after working for eight years at the Lewin Group (formerly ICF Incorporated, Lewin/ICF, and Lewin-VHI) and nearly nine years of experience in the U.S. Department of Labor. Prior to working in the Department of Labor, Dr. Barnow taught economics at the University of Pittsburgh. He has a Ph.D. degree in economics from the University of Wisconsin at Madison.

Geoff Davis is a researcher with the Signal Processing Group at Microsoft Research. Previously he was an assistant professor of mathematics at Dartmouth College. His research interests include: image compression using wavelet-based methods, adaptive signal and image representations, joint source/channel coding, and fast medical imaging acquisition. Dr. Davis has a continuing interest in labor market issues for science and engineering Ph.D.s. He is the creator of the Alfred P. Sloan Foundation sponsored website (www.phds.org) that serves as a repository of information on science career issues. He was the co-organizer of a recent CPST workshop on graduate outcomes in the sciences and engineering. He received his Ph.D. in applied mathematics form the Courant Institute of Mathematical Sciences, New York University in 1994.

Robert Dauffenbach is a professor of business administration and economics and Director, Center for Economic and Management Research, College of Business Administration, University of Oklahoma. He served on the faculties of Wayne State University and the University of Illinois prior to coming to Oklahoma. Dr. Dauffenbach's research focuses on human resource economics and quantitative methods. He has served on various study panels for the National Academy of Sciences that investigated science and engineering personnel research and data needs. He received his Ph.D. from the University of Illinois at Champaign-Urbana in 1973.

Ronald G. Ehrenberg is Vice President for Academic Programs, Planning, and Budgeting at Cornell University. In addition he serves as a Research Associate with the National Bureau of Economic Research. His research interests are concentrated in analyses of labor markets and the educational sector. Dr. Ehrenberg received his Ph.D. from Northwestern University in 1970.

Michael G. Finn is a senior economist at the Oak Ridge Institute for Science and Education, a U.S. Department of Energy facility operated by Oak Ridge Associated Universities. He worked at the Center for Human Resource Research at The Ohio State University before moving to Oak Ridge in 1976. From 1988 through 1990, Dr. Finn took leave to serve as director of studies and surveys at the National Research Council's Office of Scientific and Engineering Personnel. In all three of these jobs he has been involved in government-sponsored studies that attempt to assess future supply/demand balance of scientists and engineers. He received his Ph.D. in economics from the University of Wisconsin, Madison in 1972.

Alexander H. Flax is a consultant and is affiliated with The Washington Advisory Group. Previously he worked as an aerodynamics and structural engineer in aircraft and helicopter firms, including The Cornell Aeronautical Laboratory where he served as Vice President and Technical Director. He served as Chief Scientist of the Air Force, was appointed Assistant Secretary of the Air Force for Research and Development, and took on the additional position of Director of the National Reconnaissance Office. After leaving his government positions, Dr. Flax was President of the Institute for Defense Analyses. He is a member of the National Academy of Engineering and was its Home Secretary from 1984 to 1992. He received a B.A. in aeronautical engineering from New York University in 1940 and a Ph.D. in physics from the University of Buffalo in 1957.

Daniel S. Greenberg has specialized in coverage of science and health policy in the United States and other countries for many years. His currently a Visiting Scholar in the Johns Hopkins University Department of History of Science, Medicine and Technology, where he is writing a book on contemporary science politics. The book project is supported by a grant from the Alfred P. Sloan Foundation. He also writes a syndicated newspaper column on science policy and related matters for the *Washington Post* and other newspapers, and he is a correspondent for *The Lancet*. In addition, he is editor-at-large of *Science & Government Report*, a newsletter he founded in 1971. He formerly was a reporter on the *Washington Post*, a Congressional Fellow of the American Political Science Association, founding editor of the News and Comment Section of *Science* journal of the AAAS, and Washington correspondent of the *New England Journal of Medicine*.

Charles A. Goldman is an economist at RAND, specializing in the economics of education and international technology transfer. He is currently leading an effort to assist the White House Office of Science and Technology Policy in improving the research partnership between the federal government and higher education. This work builds on a multi-year examination of strategy and competition in U.S. higher education, with Dominic Brewer and Susan Gates. He has also completed (with William Massy) a major study of the production of science and engineering Ph.D.s in the United States. His international research examines the impact of technology and economics on U.S.-Asia relations. Prior to joining RAND, Dr. Goldman worked at the Stanford Institute for Higher Education Research and Graduate School of Business. He received his Ph.D. from the Graduate School of Business at Stanford University and also holds a B.S. in computer science and engineering from MIT.

Daniel S. Hamermesh is the Edward Everett Hale Professor of Economics at the University of Texas, Austin and research associate at the National Bureau of Economic Research. Previously he has held faculty positions at Princeton and Michigan State Universities, and has held visiting professorships at numerous universities in the United States, Europe, Australia, and Asia. He authored *Labor*

Demand, The Economics of Work and Play, and a wide variety of articles in labor economics and the leading general and specialized economics journals. Dr. Hamermesh's research concentrates on labor demand, social insurance programs (particularly unemployment insurance), and unusual applications of labor economics (e.g., suicide, sleep, and beauty). He received a Ph.D. from Yale University in 1969.

P. Brett Hammond is Director of Strategic Planning at the Teachers Insurance Annuity Association-College Retirement Equities Fund (TIAA-CREF). Prior to joining TIAA-CREF, he was Director of Academy Studies at the National Academy of Public Administration. Dr. Hammond served as Acting Executive Director and Associate Executive Director of the Commission of Behavioral and Social Sciences of the National Academy of Sciences. He was a member of the faculty of the University of California, both Berkeley and Los Angeles. Dr. Hammond's research and writings focus on pension economics, higher education economics, science and technology, finance, and policy analysis. He received his Ph.D. in public policy from the Massachusetts Institute of Technology in 1980.

Caroline M. Hoxby is an Associate Professor of Economics at Harvard University, where she has been on the faculty since 1994. In addition, she is a research fellow of the National Bureau of Economic Research. Dr. Hoxby's teaching and research are concerned with the economics of education, the labor market, and local governments. Her current research interests include a study of how market forces shape American colleges and universities. In other recent work, she has studied the growth of teachers' unionization in U.S. schools. Dr. Hoxby received her Ph.D. from MIT and also has a graduate degree in economics from Oxford, where she studied as a Rhodes Scholar.

George Johnson has served on the faculty of the University of Michigan since 1966. Previously he worked with the Department of Labor and was President of the Council of Economic Advisors. He is a specialist in labor market economics. He received a Ph.D. in economics from the University of California, Berkeley in 1966.

Nancy Kirkendall is a mathematical statistician in the Statistical Policy Office, Office of Information and Regulatory Affairs, Office of Management and Budget (OMB), and Vice President of the American Statistical Association. Previously, she worked at the Energy Information Administration, which provides forecasts of energy supply and demand as part of its mission. At OMB she is desk officer for the Census Bureau and the chair of the Federal Committee on Statistical Methodology. She is also an adjunct professor at the George Washington University, where she teaches forecasting in the Operations Research Department. She received her Ph.D. in mathematical statistics from the George Washington University in 1973.

Charles F. Manski is a Professor of Economics in the Department of Economics at the University of Wisconsin, Madison. His current research focuses on identification of problems in the social sciences. Dr. Manski received his Ph.D. from the Massachusetts Institute of Technology in 1973.

Daniel L. McFadden is Director of the Department of Economics at the University of California, Berkeley. His university experience includes E. Morris Cox Chair and Professor of Economics at the University of California, Berkeley, Fairchild Distinguished Scholar at California Institute of Technology, and Director of the Statistics Center and Economics Professor at Massachusetts Institute of Technology. Dr. McFadden's memberships include the Economic Advisory Panel of the National Science Foundation, the Executive Committee of TRB, the President of the Economet-

ric Society, the Executive Committee and Vice President of the
American Economic Association. He was the recipient of the John
Bates Clark Medal from the AEA and the Frish Medal, and is a
National Academy of Sciences member. He received his Ph.D.
from the University of Minnesota in 1962.

Ronald L. Oaxaca is a Professor of Economics in the Department of Economics and a Senior Associate in the Economic
Research Laboratory of the University of Arizona. His research
focuses on laboratory experiments with job search models and on
the accuracy and usefulness of supply estimates of scientists and
engineers. Dr. Oaxaca received a Ph.D. in economics from
Princeton University in 1971.

Cornelius J. Pings is president of the Association of American
Universities (AAU). Previously he held faculty positions at
Stanford University and California Institute of Technology and was
Provost at the University of Southern California. He is a member of
the National Academy of Engineering and has chaired the National
Academy of Sciences Committee on Science, Engineering, and
Public Policy (COSEPUP). Dr. Pings also served as a member of
the National Commission on Research, was President of the
Association of Graduate Schools, and a member of the Boards of
Directors of the Council on Government Relations and Council of
Graduate Schools. He received his Ph.D. in chemical engineering
from the California Institute of Technology in 1955.

Sherwin Rosen is the Edward L. and Betty B. Bergman Distinguished Service Professor of Economics and editor of the *Journal of
Political Economy* at the University of Chicago, Senior Fellow at the
Hoover Institute, and research associate at the National Bureau of
Economic Research. Dr. Rosen was a member of the economics
department at the University of Rochester from 1964 to 1976 before
moving to Chicago. He is known for his research contributions in

the microeconomic fields of labor economics, industrial organization, and agricultural economics on problems of product differentiation and pricing, income distribution, agricultural commodity cycles, human capital, and the economics of organizations. He received a Ph.D. in economics from the University of Chicago in 1966.

Neal H. Rosenthal is Associate Commissioner for Employment Projections and, since 1962, has worked for the Bureau of Labor Statistics on programs concerned with developing current and projected occupational employment. He is the author of numerous articles and reports dealing with current and projected supply and demand in a variety of occupational fields. He currently directs the Bureau's work on developing national labor force, industry employment, and occupational employment projections and career guidance information, including the *Occupational Outlook Handbook* and *Occupational Outlook Quarterly*.

Jack H. Schuster is professor of education and public policy at the Claremont Graduate School. Previously he served as Assistant Director of Admissions at Tulane University, Assistant to the Chancellor at the University of California, Berkeley, and as a lecturer in political science. He has held the positions of Visiting Scholar at the University of Michigan's Center for the Study of Higher Education, Guest Scholar in the Governmental Studies Program at the Brookings Institution, and Visiting Fellow and Research Associate at the University of Oxford. Dr. Schuster's research activities are centered on the academic labor market and accreditation activities. He received a J.D. degree from Harvard Law School and a Ph.D. from the University of California, Berkeley.

Paula E. Stephan is Associate Dean of the School of Policy Studies at Georgia State University. Her research is centered on the distribution of rewards to scientific research, an examination of

scientists as entrepreneurs, and variations in labor supply.
Dr. Stephan received a Ph.D. in economics from the University of Michigan in 1971.

William A. (Skip) Stiles, Jr. is the Democratic Legislative Director for the House Committee on Science, a position he has held since 1991. Mr. Stiles' current job responsibilities include working with the democratic congressional leadership and the committee's democratic members to coordinate the legislative agenda. His specific responsibilities have included: lead committee staffer for the Energy Policy Act of 1992, work on defense conversion technology programs, environmental technology legislation, and alternative transportation technologies. Prior to his current position, Mr. Stiles was staff director for the Department Operations, Research, and Foreign Agriculture Subcommittee on the House Agriculture Committee, a position he held since 1985. Mr. Stiles is a 1971 graduate of the College of William and Mary.

Sarah Turner is an assistant professor of education and economics at the University of Virginia. Previously she worked for the Mellon Foundation. Her recent research focuses on the impact of changes in financial aid programs on institutional tuition and aid policies, gender differences in the choice of major and occupational outcomes, the assessment of changes in the return to college quality over time, and the relationship between schools of education and the labor market for teachers. She received her B.A. from Princeton and did her doctoral study in economics at the University of Michigan.

George E. Walker is Vice President of Research and Dean of the Graduate School at Indiana University, Bloomington. Previously he was a research associate in physics at Los Alamos Scientific Laboratory, a research associate at Stanford University, and a member of the physics faculty at Indiana University, Bloomington.

He serves as a visiting staff member at Los Alamos Scientific Laboratory. Dr. Walker's research activities are focused on nuclear theory, electron scattering, meson-nucleus interactions, nucleon-nucleus interactions, and heavy ion scattering. He received his Ph.D. in physics from Case Western Reserve University in 1966.

Eric Weinstein is currently conducting postdoctoral research at the Massachusetts Institute of Technology. Previously he served as a Lady Davis Fellow at the Hebrew University in Jerusalem. In his current research he is collaborating with Pia Malaney of the Harvard Institute for International Development on a program which uses geometry and mathematical models to solve problems in inter-temporal economics (e.g., changing preferences and index number problems). Dr. Weinstein is continuing his analysis of scientific and engineering supply and demand theory, and is beginning collaboration with Richard Freeman on a framework for analyzing the scientific labor market. He received his Ph.D. from Harvard University.

WORKSHOP PARTICIPANTS

March 19-20, 1998
National Academy of Sciences
Washington, D.C.

John Armstrong
NAE
Amherst, MA

Tom Arrison
Policy Division
National Research Council
Washington, DC 20418

Burt Barnow
Institute for Policy Studies
Johns Hopkins University
Baltimore, MD

Kenneth Brown
Social, Behavioral and Economic Sciences
National Science Foundation

Bill Colglazier
National Research Council Executive Office
Washington, DC 20418

Roman Czujko
Statistics Division
American Institute of Physics
College Park, MD 20740-3843

Robert Dauffenbach
Center for Economic Management Research
University of Oklahoma
Norman, OK 73019

Geoff Davis
Department of Mathematics
Dartmouth College
Hanover, NH 03755

Ronald Ehrenberg
Academic Programs, Planning, and Budgeting
Cornell University
Ithaca, NY 14853-2801

Michael Finn
Oak Ridge Institute for Science and Education
Oak Ridge, TN 37830

Alexander Flax
National Academy of Engineering
Potomac, MD

Catherine Gaddy
Commission on Professionals in Science and Technology
Washington, DC 20005

Charles Goldman
RAND
Santa Monica, CA 90401

Mary Golladay
Science and Engineering Education and Human Resources
National Science Foundation
Arlington, VA 22230

Daniel Greenberg
Department of History of Science, Medicine and Technology
Johns Hopkins University
Baltimore, MD

Jeanne Griffith
Science Resources Studies
National Science Foundation
Arlington, VA 22230

Daniel Hamermesh
Department of Economics
University of Texas
Austin, TX 78712-1173

Brett Hammond
Research Division
TIAA/CREF
New York, NY 10017

Dan Hecker
Bureau of Labor Statistics
Washington, DC 20212

Peter Henderson
Office of Scientific and Engineering Personnel
National Research Council
Washington, DC 20418

Caroline Hoxby
Department of Economics
Harvard University
Cambridge, MA 02138

George Johnson
Economics Department
University of Michigan
Ann Arbor, MI 48109-1220

Mary Jordan
American Chemical Society
Washington, DC 20036

Nancy Kirkendall
Office of Management and Budget
Washington, DC 20503

Charlotte Kuh
Office of Scientific and Engineering Personnel
National Research Council
Washington, DC 20418

Carrie Langer
Policy Division
National Research Council
Washington, DC 20418

Al Lazen
Commission on Life Sciences
National Research Council
Washington, DC 20418

Rolf Lehming
Social, Behavioral and Economic Research
National Science Foundation
Arlington, VA 22230

Steve Lukasik
OSEP Advisory Committee
Los Angeles, CA 90077

Charles Manski
Department of Economics
Northwestern University
Evanston, IL 60208

Daniel McFadden
Department of Economics
University of California
Berkeley, CA 94720-3880

Carol Ann Meares
Office of Technology Policy
U.S. Department of Commerce
Washington, DC 20230

Steve Merrill
Policy Division
National Research Council
Washington, DC 20418

Michael Neuschatz
American Institute of Physics
College Park, MD 20740

Ronald Oaxaca
Department of Economics
University of Arizona
Tucson, AZ 85721

Leigh Ann Pennington
Oak Ridge Institute for Science and Education
Oak Ridge, TN 37830

Cornelius Pings
Association of American Universities
Washington, DC 20005

Roy Radner
Stern School of Business
New York University
New York, NY 10012

Alan Rapoport
Science Resources Studies
National Science Foundation
Arlington, VA 22230

Mark Regets
Science Resources Studies
National Science Foundation
Arlington, VA 22230

George Reinhart
Office of Scientific and Engineering Personnel
National Research Council
Washington, DC 20418

Sherwin Rosen
Department of Economics
University of Chicago
Chicago, IL 60637

Neal Rosenthal
Office of Employment Projections
Bureau of Labor Statistics
Washington, DC 20212

Nicole Ruediger
Science's New Wave
AAAS
Washington, DC 20005

Walter Schaffer
Research Training & Special Programs
National Institutes of Health
Bethesda, MD 20892-7910

Jack Schuster
Center for Educational Studies
Claremont Graduate University
Claremont, CA 91711-6160

Paula Stephan
School of Policy Studies
Georgia State University
Atlanta, GA 30303-3083

Skip Stiles
Committee on Science
U.S. House of Representatives
Washington, DC 20515

Marilyn Suiter
American Geological Institute
Alexandria, VA 22302

Jennifer Sutton
Office of Scientific and Engineering Personnel
National Research Council
Washington, DC 20418

Peter Syverson
Research and Information Services
Council of Graduate Schools
Washington, DC 20036

Molla Teclemariam
Office of Scientific and Engineering Personnel
National Research Council
Washington, DC 20418

Michael Teitelbaum
Alfred P. Sloan Foundation
New York, NY 10111-0242

Eleanor Thomas
Office of Policy Support
National Science Foundation
Arlington, VA 22230

Alan Tupek
Science Resources Studies
National Science Foundation
Arlington, VA 22230

Sarah Turner
Curry School of Education
University of Virginia
Charlottesville, VA 22903-2495

Jim Voytuk
Office of Scientific and Engineering Personnel
National Research Council
Washington, DC 20418

George Walker
Graduate School
Indiana University
Bloomington, IN 47405-3901

Katherine Wallman
Office of Management and Budget
Washington, DC 20503

Eric Weinstein
Department of Mathematics
Massachusetts Institute of Technology
Cambridge, MA 02139

John Wiley
OSEP Advisory Committee
University of Wisconsin
Madison, WI 53706

Steven Williams
American Psychological Association
Washington, DC 20002-4242